Paul Sorauer, Frederick Ernest Weiss

A Popular Treatise on the Physiology of Plants

For the use of gardeners or for students of horticulture and of agriculture

Paul Sorauer, Frederick Ernest Weiss

A Popular Treatise on the Physiology of Plants
For the use of gardeners or for students of horticulture and of agriculture

ISBN/EAN: 9783337082734

Printed in Europe, USA, Canada, Australia, Japan

Cover: Foto ©berggeist007 / pixelio.de

More available books at **www.hansebooks.com**

Printed by BALLANTYNE, HANSON & CO.
At the Ballantyne Press

TRANSLATOR'S PREFACE

In undertaking a course of lectures on Vegetable Physiology at the Royal Botanical Gardens in Manchester, I felt very much the want of a book which I could place in the hands of the gardeners and students of horticulture who attended the course, and which, while giving them a thoroughly scientific account of the functions of the various organs, would at the same time deal with the practical applications of the principles of Vegetable Physiology.

Such a book was accessible to me in the German in Professor Sorauer's admirable " Populäre Pflanzenphysiologie," and I should have been thankful if, at the time, a translation had been available for my hearers. Professor Sorauer is well known as an authority on the diseases of plants, but he had an additional qualification for writing his book, namely, that he was for many years the director of an institution which had in view the scientific training of gardeners and agriculturists.

We are not so fortunate in this country as to possess such well-appointed and State-endowed experimental stations as those which exist in Germany, and we must for the present be content with the more humble ventures instituted by the County Councils. In all such educational courses as they may institute, it is to be hoped the scientific principles which underlie the practice of horticulture and of agriculture will not be lost sight of, and those intrusted with the teaching of these principles will do well to follow the lines laid down in this treatise by Professor Sorauer, and to discuss the often very direct bearing of the physiological truths upon the questions which exercise the minds of gardeners and agriculturists. For these the book is directly written, though other students of botany will find much useful information in this treatise, emphasis being laid by the writer upon the fundamental truths by a repeated treatment of them under various aspects.

In conclusion, I would thank Dr. Maxwell T. Masters for information on some of the technicalities of horticulture, always given so freely and so kindly.

F. E. WEISS.

TABLE OF CONTENTS

CHAPTER I

INTRODUCTION

SEC.		PAGE
1.	What conception should a gardener form of a vegetable organism?	1
2.	What are the functions of the various organs of a plant?	2

CHAPTER II

THE STRUCTURE OF THE ROOT

3.	What is the structure of the absorptive organ, the root?	4
	(*a.*) The absorptive portion of the root.	4
	(*b.*) The root-tip	7
	(*c.*) The conducting portion of the root	10
	(*d.*) The process of conduction	16

CHAPTER III

THE NUTRITION OF THE ROOT

4.	What substances must be present in the soil for the continuous and effective nutrition of plants?	30
5.	What is the effect of the various nutritive substances on the plant, and in what form do they enter it?	31
6.	What part is played by the subsidiary nutritive substances and the occasional admixtures to the nutritive salts?	35
7.	In what form does the root find the nutritive substances in the soil?	39
	(*a.*) Potassium	41
	(*b.*) Phosphoric acid	42
	(*c.*) Nitrogen	42

CONTENTS

SEC.		PAGE
8.	Why and how should we replace the chief nutritive substance in the soil?	52
	(a.) Fallow	52
	(b.) Inorganic manures	53
	(c.) Organic manures	55
	(d.) Stable manures	56
9.	How can the soil best meet the requirements of the roots for air?	61
10.	How can we improve our fields so as to obtain the best possible crops?	65
11.	How is the nutrition of pot-plants effected?	70
12.	How do aërial roots nourish a plant?	75
13.	How do ordinary roots obtain their necessary supply of air?	77

CHAPTER IV

THE TREATMENT OF ROOTS

14.	How should roots be treated in transplanting?	81
	(a.) A root system which has been considerably pruned	81
	(b.) A root system from which only the delicate roots have been removed	85
	(c.) The treatment of roots in re-potting	85
	(d.) The treatment of roots in transplanting in the open	87

CHAPTER V

THE STEM

15.	What is the structure of the stem?	94
16.	What is cambium?	98
17.	What is the function of the cambium in the ordinary course of growth?	103
18.	How does the stem as a whole perform its functions?	105

CHAPTER VI

THE LEAF

19.	Which cells of the leaf are the most essential?	108
20.	How are the assimilating cells of the leaf protected?	110
21.	How are the assimilating cells arranged within the leaf?	118
22.	How is the leaf developed?	120
23.	What substances does the leaf chiefly form?	122
24.	How does the leaf actually perform its assimilatory function?	127

CHAPTER VII

THE TREATMENT OF THE SHOOT

SEC.		PAGE
25.	Why must the shoots of our cultivated plants be pruned?	134
26.	What is the least injurious form of a cut?	135
27.	How does summer pruning differ in its effects from winter pruning?	137
28.	What is the effect of the different methods of pruning?	140
29.	How may pruning be used to regulate the natural development of the tree?	142
30.	When is pruning harmful?	146
31.	In what way can the natural process of healing be accelerated?	149
32.	By what means can we increase the effect of pruning?	155
	(*a.*) The bending of shoots	155
	(*b.*) The twisting of shoots	157
	(*c.*) Notching	158
	(*d.*) Ringing	159
	(*e.*) Peeling the stems	161
33.	Why do we slit the bark of trees?	164

CHAPTER VIII

THE USE OF SHOOTS FOR PROPAGATING

34.	What is meant by layering, and of what use is it?	169
35.	What rules should be followed in striking cuttings?	172
36.	What objects have we in view in budding and grafting, and how are these operations best performed?	183
37.	To what extent do scion and stock mutually influence one another?	194

CHAPTER IX

THE TREATMENT OF LEAVES

38.	What is the effect of injuries to the leafy tissues?	199
39.	In what cases can the leaf be used for propagation?	201

CHAPTER X

THE THEORY OF WATERING

40.	Why must we pay special attention to the watering of plants?	205

CHAPTER XI

THE FLOWER

SEC.		PAGE
41.	Of what parts does the flower consist? . . .	214
42.	How are single and how are double flowers developed?	218
43.	Can a gardener determine the development of flowers?	221

CHAPTER XII

FRUITS AND SEEDS

44.	How are fruits and seeds formed?	226
45.	How can the formation of fruit be influenced by different methods of cultivation?	233
46.	What are the conditions governing the production of seeds?	238
47.	How should the ripe seed be treated?	244

INDEX . 253

THE PHYSIOLOGY OF PLANTS

CHAPTER I

INTRODUCTION

§ 1. What conception should a gardener form of a vegetable organism?

IT is the business of the gardener and of the horticulturist to cultivate plants. This term implies the guiding of the natural development of a plant towards some special end. For a gardener ought to know how to make use of the natural processes and of the development of the vegetable organism in such a way as to realise as completely as possible the end which his cultivation has in view. His endeavours may be of various kinds. In many cases he only tries to cultivate a plant in its normal and natural form in some special locality. This is usually the case in landscape-gardening or in the cultivation of plants from warmer regions in our own climate. In other cases, which we might call "true cultivation," the gardener seeks to make a certain plant more useful for his purposes by increasing its productivity and improving the quality of its fruits. This is clearly the case in the culture of those plants which are grown as vegetables, in which case, those organs which are used as food are increased in size and number, and are considerably improved in delicacy. Here the gardener interferes with the normal development of a certain species, so as to increase certain functions of the plant, and thus causes an increased development of certain organs. In other cases, the time of flowering or fruiting of a certain species is changed; as, for instance, when we wish to obtain spring flowers in winter (*forcing*), or, on the contrary, when we wish to produce spring flowers late in the summer.

We are able, also, by certain methods of cultivation, to change the colour of flowers, or to cause them to double by a transformation of stamens into petals, or by the formation of new whorls of leaves. We can cause hard roots to become thick and succulent, as in the beets, and hard fruits to become large and juicy by an increase of their softer tissues (apples and pears); and we may therefore consider ourselves able to a certain degree to modify the characters of the vegetable organisms, and to partially change their normal development.

A plant must not, therefore, be looked upon as an unchangeable organism, restricted to a definite form, but as a plastic organism, capable of further modification in all its parts. Its usual shape can be altered as if it were made of wax, and it can be remodelled within certain limits. This remodelling of the form of a plant is, however, only possible if the gardener understands how to regulate the conditions of its life, so that without damaging the whole, *i.e.*, the life of the plant, the functions of the various organs may be increased beyond the normal amount or reduced below the usual limit in favour of some other organ. To effect these changes, it is naturally essential to possess an insight into the various processes which make up the life of a plant, and to understand how they are affected by various external conditions—in fact, to possess a knowledge of vegetable physiology.

But though our knowledge of vegetable physiology is by no means complete, still we have obtained some insight into many of the processes of plant-life, and a gardener must learn how to make practical use of this knowledge. For this reason we start by asking:

§ 2. What are the functions of the various organs of a plant?

A plant presents itself to us as a complex organism, built up of a number of chambers or cells, each of which has its special and, to a certain extent, independent function. The functions of all cells of the body have two common objects in view, namely, the preservation of the individual and the production of material which may be employed in the formation of new individuals. These latter are formed in the seeds, the function of which is to perpetuate the species.

Of the various members of the vegetable organism, the flowers have the function of producing and developing new individuals or offspring from the organic material with which they are provided. The leaves, on the other hand, are concerned in the production of the necessary food material. This is effected in the green cells of the leaves under the influence of light and from the raw material, which they obtain partly from the atmosphere (carbonic acid), and partly from the soil in the form of mineral salts dissolved in water. It is the function of the roots to fix the plant in the soil, and to absorb from it the soluble mineral salts.

CHAPTER II

THE STRUCTURE OF THE ROOT

§ 3. What is the structure of the absorptive organ, the root?

WE are first of all interested in the general structure of the root. This organ represents the direct downward continuation of the stem, and presents, like the latter, a central cylinder (Fig. 1 G), which is surrounded by a softer tissue, the cortex (Fig. 1 R). As the woody cylinder of the uninjured root never reaches the periphery, but is always surrounded by cortical tissue, it is necessarily the latter which is especially concerned in the absorption of nutritive substances from the soil. The central woody tissue, in which the naked eye can often distinguish large wide vessels (G), has the function of conducting the dissolved substances to all parts where they may be required.

(a.) The Absorptive Portion of the Root.

If we take a healthy plant carefully from loose soil, so as not to injure the delicate ramifications of the root, and if we wash the latter under a moderately strong jet of water, it will be possible to free the roots almost entirely from the adhering particles of soil. Only one portion of each rootlet will retain its covering of grains of sand or particles of earth, and these will seem to be stuck on to the root. With the aid of a microscope we observe that delicate contorted root-hairs (Fig. 1 H) have in reality fixed themselves to these small particles of the soil and hold them very tightly. That region of the root which is provided with this covering of sand is the portion which is able to take up water from the soil. The absorptive region of the root is therefore restricted to a certain youthful zone, which begins close behind the smooth, often transparent root-tip and reaches back about an inch. The older, brown portions of the root are useless for the purpose of absorption.

If we examine such a root-hair from its tip to its point of origin, we find it to be a slender tube filled with liquid, and projecting like the finger of a glove from one of the external cells of the root. These cells, which form the outer layer (*epidermis*) (Fig. 1 E) of the root, and which are very thin-walled in this region, have, by elongating to form root-hairs, immensely increased their absorptive surface. The root-hairs, therefore, enable the root to present in a very small region an enormously large absorptive area.

As the chief substance which is to be absorbed by the root is water, we find that the development of the absorptive surface of the root is often directly proportional to the need for water. In the case of aquatic plants, for instance, in which the entire plant is constantly in contact with water, there is no need for an increase of the absorptive surface, and the production of root-hairs seems unnecessary. As a matter of fact, the roots of these plants are often destitute of root-hairs. The roots of the water-hemlock and other plants have been observed to be smooth as long as they live in water, but as soon as the plants are grown in dry soil, their roots become clothed with root-hairs. In typical land plants, on the other hand, the necessity for an increase of the absorptive surface will become greater the larger the leaves are, and the greater therefore the transpiratory surface becomes. Plants, on the other hand, possessing narrow, tough leaves, which reduce the transpiration, will be able to absorb sufficient water even if no root-hairs are developed. This indeed is the case, and the roots of various Conifers which bear needle-shaped leaves have been found to be devoid of root-hairs, the thin flattened layer of epidermal cells being able to absorb sufficient water for the needs of the tree. Sometimes on the same plant some roots may be devoid of hairs, while others are sparingly, others largely covered with root-hairs, and this will depend upon the facility which the roots experience in obtaining their water supply. The more difficult they find it to obtain water, the more perfectly will the absorbing hairs be developed. In perfectly dry soil, too, the development of root-hairs will naturally be stopped, as the root will scarcely grow at all. The development of root-hairs is most abundant in a damp atmosphere.

We have already mentioned that the particles of soil cannot be removed from the absorbing portion of the root, but adhere to it pretty firmly. If seedlings are grown in a damp atmosphere on moderately damp sand, in a short time many root-hairs will be found which have come in contact with grains of sand. Where this is the case, they will flatten themselves against the particle, will curve themselves round it, and may grow on to repeat the process with another grain of sand. The cessation of growth in length and the expansion of the point of contact are not, however, the only phenomena which make their appearance when a root-hair touches a particle of soil. The outer wall of the root-hair will be seen to lose its sharp outline and swell up into a slimy mass. It is this mucilaginous coating which fastens the grains of sand so firmly to the root-hairs. Now let us imagine that instead of a grain of sand, the root-hair has grown round a particle of soil which is covered with a thin deposit of the salts necessary for the plant. As a matter of fact, the mineral salts which are most essential to plant life are absorbed by the soil, and are mechanically held by each particle of soil, together with a thin pellicle of water which surrounds each particle. Now when a root-hair comes into close contact with such a particle, the mucilage will help to dissolve the salts, and the latter pass through invisible pores of the membrane or cell-wall into the root-hair (see Fig. 1 H), saturate the liquid or sap it contains, and permeate the membrane which separates it from the next cell within, belonging to the region R. If this inner cell of the cortex is in want of more nutritive matter, it will absorb some from the root-hair and satisfy its own wants. But that will not conclude the matter, for the neighbouring cells will in their turn absorb from the cortical cell, only, however, to pass on these substances to still more eager cells which lie above them. In this way a passage of the nutritive substances takes place towards the upper regions of the plant, towards the growing points, which are always in want of food material, and as these substances are constantly being used up in these regions, the lower cells, from which they have been withdrawn in their turn, absorb from still lower cells, and these ultimately from the root-hairs. Thus a continuous current of nutritive salts is established through the cell-walls, beginning in the root-hairs and extending to the vessels which occupy the central cylinder of the root, and through which they are forced with great rapidity to the upper regions of the plant.

But even solid mineral particles can be made use of by the root-hairs. For these cells give off carbonic acid and also non-volatile acids, which act upon the particles, dissolve them, and absorb the solution. This process can be very readily and effectively demonstrated by growing seedlings under a very thin covering of soil on a polished piece of marble. After a short time those rootlets which have come in contact with the marble plate will be found to have etched their course over its surface. This they will have done by dissolving away the marble from those places where they have been in contact with it.

(b.) *The Root-Tip.*

It is of great importance to the gardener who wishes to regulate the growth of roots to suit his own purposes, to understand the structure of that part of the root in which growth takes place. The latter depends, first, upon the formation of new root-cells; secondly, upon the elongation of these cells.

We should, however, be making a mistake were we to assume that the formation of the tissues which cause the elongation of the root takes place in the same way as it does in the case of the shoot. In the latter the tip, at first protected by bud-scales, is afterwards generally uncovered, and is built up of very delicate, closely packed cells, rich in protoplasm and divided by new transverse walls. The extreme tip of this actively growing region (*meristem*) is continuously giving rise to new cells during the whole vegetative period, and the increase in length of the axis is due to the growth of each of the new cells to its mature size.

In the case of the root, too, we find, it is true, the same method of arrangement, and increase of the meristematic

8 THE PHYSIOLOGY OF PLANTS

cells of the root-tip, and the increase in length is therefore also due to apical growth. But this root-tip is not free; it is covered over by a cap of parenchymatous cells called the root-

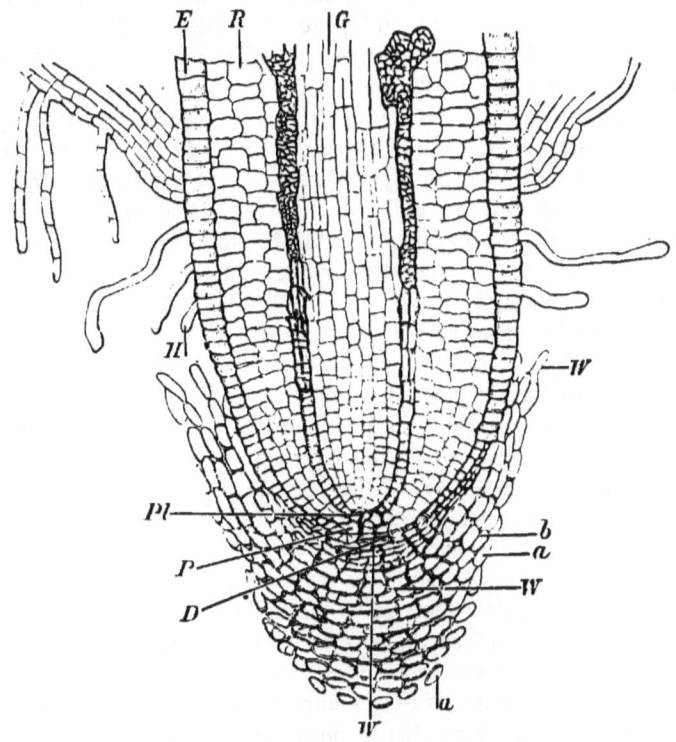

FIG. 1.—LATERAL ROOT GROWING OUT FROM THE STEM OF A POTATO.

G young central cylinder, bounded by rows of wood vessels; R, cortex; E, epidermis; H, hairs. The separation of these tissues can be traced into the youngest portions of the root-tip in this meristematic region. Pl (Plerom) is the tissue which will continue the formation of the central cylinder; P (Periblem) the seat of formation of the cortex, and D (Dermatogen) the initial cells of the epidermis. The delicate meristematic apex of the root itself is covered by the root-cap (W), the outermost cells of which (a) are gradually becoming disorganised while internally (within b) new root-cap cells are being formed.

cap (Fig. 1 WW). The layers of cells forming the latter are so arranged that the oldest layers (a) lie nearest the outside, the youngest layers (b) next to the root-tip; here the root-cap is connected with the root-tip by a layer of cells common to both.

If we imagine a finger covered by a thimble, which adheres to the finger by its internal surface, we have a rough but pretty accurate illustration of a root-apex with root-cap.

Beneath this root-cap lie the tissues of the root-tip, and in different arrangement according to the different genera. In the majority of cases a central group of meristem cells can be distinguished (Fig. 1 *Pl*), which give rise to the new portions of the central vascular cylinder. This central portion is surrounded by a number of layers of cells (*P*) which give rise to the cortex. A single layer of cells (*D*) covers in the cortex over its whole extent, and will form the epidermis and the absorbing cells or root-hairs. At the very extreme point of the root-tip these tissues all converge into a group of cells, which is the seat of formation of all these layers, and is also the point of connection between the root-tip and the root-cap. The root itself, therefore, and the root-cap have in reality the same point of origin, namely, the extreme apex of the root; and in the generality of cases, in dicotyledonous plants, we may look upon the root-cap as merely a production or proliferation of the youngest epidermal layer of the root itself.

This protuberance of the root is renewed from its inner layer, *i.e.*, from its point of connection with the root itself, while its outer and therefore oldest cells (*a*) gradually lose their contents, allow their cell-wall to break down into mucilage, and in the end separate from each other and break down. During this process carbonic acid is given off, and becoming absorbed by the water, it causes mineral salts, such as phosphate of lime, which are insoluble in pure water, to become dissolved.

This we may look upon as one of the functions of the root-cap. The salts which the decay of the older root-cap cells causes to be dissolved are absorbed by the inner younger cells and conducted to the centres of growth. To that extent, therefore, the root-cap may be considered an organ of absorption; its chief function, however, is that of a protective cap. We can readily understand what obstacles a continuously elongating root-tip has to encounter in its efforts to force a passage through the firm soil. It requires the application of special forces and special means of protection. The force which enables the root to bore its way into the soil is the elongation, already alluded to, of the cells lying behind the root-apex, which have lost their meristematic character and have become transformed into the permanent tissues of the root. By their growth in length they push the apex forward, as the older portion of the root is firmly fixed in the soil by the root-hairs, and cannot therefore be forced back. But how could a soft root-tip, composed as it is of the most delicate cells, be forced between the sometimes very sharp particles of the soil without sustaining severe injury? Nature has guarded against any such damage by protecting the root with a cap, the oldest cells of which are turned towards the outside, and protect the growing cells at the interior from any harm.

In the case of other organs (leaves, for instance) we can see how their peculiar development is an adaptation to the surrounding conditions, and we might presuppose the same for roots; nor shall we go far wrong if we assume that the production of the root-cap is a necessary adaptation of the root to its conditions of growth, and that it has thus gradually been evolved.

The centre of growth, the elongating and the absorbing regions of the root, are all confined to the root-tip, and consequently the conclusion we arrive at from our preceding observations is that **the first condition of a rational method of cultivation is to preserve and increase the number of root-tips.**

(c.) *The Conducting Portion of the Root.*

After having examined the seat of growth in the root, and having determined the region of absorption, it is necessary to look for the tissue through which the absorbed water with its dissolved food substances is conducted. This tissue is the **woody** tissue of older portions of the root, which, as can easily be seen in any root of a tree, forms a hard central cylinder, and is surrounded by a soft cortical layer. To form the central cylinder, the elements which we have mentioned as G in Fig. 1 become more and more numerous, and group themselves into thick strands arranged in a ring. These strands are termed the "**vascular bundles.**" In both Mono- and Dicotyledons they consist of various elements, some of which are of the nature of narrow or wider tubes (*vessels*), while others take the form of long, narrow, and very much thickened fibres (*wood cells*).

The vessels are formed of superposed cells, which have become joined by the loss of their transverse walls, and thus form continuous canals. The lateral walls become thickened by the addition of woody substance on their inner surface. This addition of substance to the original cell-wall (*primary membrane*), which only becomes visible as the vessels become old, constitutes the secondary membrane, and appears sometimes as a fairly continuous thickening, leaving certain thinner places, the pores or pits (*pitted vessel*, Fig. 2 *g*). In some cases the thickening is netlike (*reticulate*), ladderlike (*scalariform*), or presents the appearance of a spiral band on the inside of the primary wall (Fig. 2 *g′*). According to the manner in which

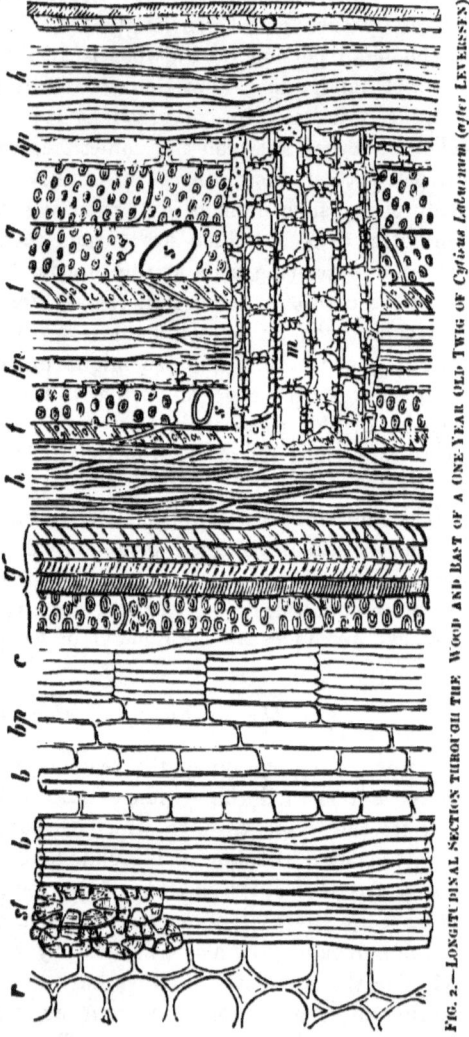

FIG. 2.—LONGITUDINAL SECTION THROUGH THE WOOD AND BAST OF A ONE YEAR OLD TWIG OF *Cytisus Laburnum* (*after* LEUNIS).

the thickening takes place we distinguish reticulate, scalariform, spiral, or annular vessels. The fibrous elements which run between the vessels and give the bundle its rigidity are shown in Fig. 2 *h*. They are wood cells which have become wedged together by their pointed ends, and which, by a chemical change of their wall (*lignification*) have attained a great hardness, but at the same time have become very brittle. Their great length, thickness, and pointed ends distinguish them from those cells which build up the soft tissues of herbaceous plants, and which are usually six-sided, thin-walled, and richer in contents. These latter are termed **parenchymatous cells** (Fig. 1 *R*, and Fig. 2 *r*), while the former constitute the **prosenchymatous tissues.**

The hard bast cells (Fig. 2 *b*) are a special form of prosenchyma; they are usually much longer than the wood cells, and often much more thick walled, but they are flexible and tough, and run on the outside of the woody cylinder, but parallel to it. The bast used so largely in gardening operations consists almost entirely of such cells. These hard bast cells generally form broad longitudinal bands or strips. Protected by these cells, and between them and the woody cylinder, we find long thin-walled cells, rich in protoplasm or tube-like cells, which have this peculiarity, that the transverse walls separating one from the other are perforated by pores, and they have therefore received the name of **sieve-tubes.** These tubes are accompanied by thin-walled, bluntly ending cells, the **bast parenchyma** (Fig. 2 *bp*). In like manner we find between the wood vessels and wood fibres parenchymatous cells, which have, however, also become lignified; they represent the **wood parenchyma** (Fig. 2 *hp*).

The vascular bundles which we find in plants running along both the stem and the root, and branching into the leaves to form a network of veins, consist therefore of two very distinct portions. The most largely represented is the wood, consisting of brittle lignified elements (Fig. 2, from *g′* to *hp*, and *h*). The wood consists of vessels, narrow wood fibres (*libriform*), and wood parenchyma. In the case of Conifers the first mentioned are replaced by wide vessel-like cells, which are, however, not continuous, but resemble the wood fibres. On account of their similarity with the vessels, however, which are sometimes called **tracheæ**, they have been termed **tracheids**, and they are further characterised by a curious form of pores termed **bordered pits**. The tracheids perform the same function as the vessels of other plants. They conduct the nutritive salts dissolved in water from the soil first into the leaves, where, under the influence of the light and together with the carbonic acid of the air, they are transformed into organic food matter. The other portion of the fibro-vascular bundle lies towards the outside, and consists of flexible, thin-walled sieve-tubes and cambiform cells rich in protoplasm. This soft portion of the bundle, which is concerned in the conduction of the organic nitrogenous food substance, and is termed the **soft bast**, is generally protected on the outside by strands of thick-walled **hard bast** cells (*b*).

These bundles, whether they consist of real vessels or tracheids, conduct along their woody portion the raw materials which the roots have absorbed, and carry them into the finest ramifications of the stem. Through their soft bast tubes, on the other hand, they conduct the elaborated substances which are necessary for the increase in root and shoot from the leaves, where they are formed, to the root-tips. As these conducting cells are very delicate, and might easily be compressed and damaged by any external pressure, we find them generally protected on the outside by a band of hard bast cells.[1]

Let us now examine also in cross-section the structure of the root which we have studied from a longitudinal section. We choose as an example the root of a monocotyledonous plant, namely, that of the Shallot (*Allium ascalonicum*) (Fig. 3), and a root of a dicotyledonous plant, that of the Auricula (*Primula Auricula*) (Fig. 4). In both figures the soft cortical tissues, which are of secondary importance in the present consideration, as well as the epidermis and root-hairs, have been left out, and are only indicated by a few layers of cells at the

---

[1] These protective strands of hard bast have been considered as part of the vascular bundle, and the hard and soft bast together have received the name of *phloëm*, while the wood vessels together with the wood fibres have been called the *xylem*. Both parts together form the *fibro-vascular bundle*.

outside. The chief conducting apparatus for the food material which has been absorbed by the root-hairs and passed on through the cortical tissue is the central cylinder, which contains wood vessels, indicated in Fig. 1 *G* by narrow rows of cells. These vessels, which appear in Figs. 3 and 4 as darkly outlined spaces, do not lie singly in the transverse sections, but are collected in groups. In the root of the Shallot, as in many other Monocotyledons, the groups of vessels are arranged

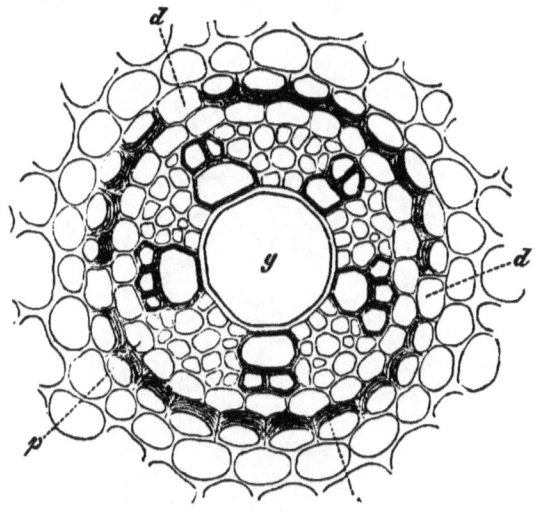

FIG. 3.—THE CENTRAL VASCULAR CYLINDER OF THE ROOT OF THE SHALLOT
(*Allium ascalonicum*).

g is the very wide central vessel which joins the five dark rays of smaller wood vessels; *p* is the layer of cells which gives rise to new lateral roots (*pericambium*); *s*, the endodermis or bundle-sheath; *d*, the thin passage cells (*after* HABERLANDT).

in rays, or like the spokes of a wheel round a central very large vessel (Fig. 3 *g*), so that we seem to have a single group of wood vessels running out into five arms or plates (Figs. 3, 4). In the root of the Auricula (Fig. 4) the seven separate groups of wood vessels (*g*) are arranged in a ring round a soft parenchymatous central tissue, which is termed pith or medulla. In the Dicotyledons this pith often replaces the large central vessel of the Monocotyledons, and the root structure resembles more closely the structure of the stem. We need only imagine that the groups of wood vessels (Fig. 4 *g*) have become enlarged by the addition of new vessels and wood cells, so as to touch each other laterally, and then we should have a continuous cylinder of wood such as we find in any first year's twig.

Between the dark groups of wood vessels we see in Fig. 4 *s* lighter groups of smaller cells. These are the more delicate passages of the soft bast (*leptom*), which, we have previously mentioned, conduct through their sieve-like transverse walls

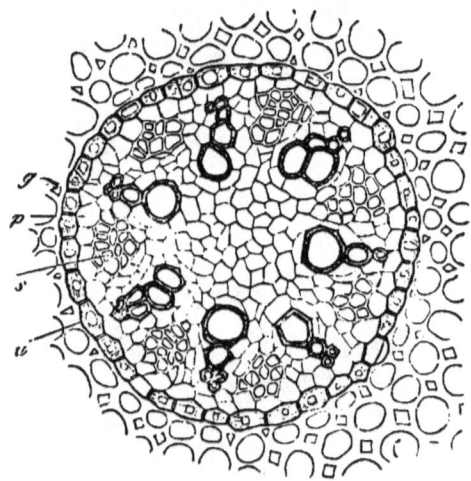

FIG. 4.—SEVEN-RAYED VASCULAR CYLINDER OF AN ADVENTITIOUS ROOT OF
Primula Auricula.
g, wood rays ; *s*, groups of soft bast cells ; *p*, pericambium ; *u*, bundle-sheath (*after* De Bary).

the plastic food substances formed in the leaves to the root system, for the furtherance of its growth. The tissues for the upward passage of the raw mineral salts which are absorbed by the roots lie at the side of the tubes, which have to conduct the formed substances down from the leaves. Both tissues are surrounded by a ring of lightly coloured cells, *p*, the **pericambium**. It is in this layer that the first rudiments of the new lateral roots arise, and in such places the surrounding ring of thick-walled cells which protects all the conducting tissues is interrupted. This ring (Fig. 3 *s*, and Fig. 4 *u*) is termed the **bundle-sheath** (*endodermis*), and when the soft cortical tissues of the roots are destroyed, it keeps the central cylinder intact.

As a matter of fact, the cortex of the root of many grasses and other Monocotyledons dies off at an early period, and then the root continues to live and function protected by the endodermis alone.

But a matter of great importance in the structure of the root of *Allium ascalonicum* (Fig. 3), and one deserving special mention, is the fact that the interruption of this protective sheath by thin-walled cells (Fig. 3 *d*) always takes place immediately opposite a group of wood vessels.

For if we remember that these thin-walled cells offer an easy passage for watery fluids, while the thick-walled cells are rendered impervious by their thick cell-walls, we see that these cells (*d*) constitute the points of passage of the water which has been taken up by the root-hairs, and, after traversing the cortex, has to make its way into the wood vessels, and through them up into the stem and its branches.

(*d*.) *The Process of Conduction.*

Having recognised the thin-walled cells of the endodermis as passages through the thick-walled protective sheath of the root, we have gained a sufficient insight into the structure of the root, and can now pass on to the consideration of the physical forces which must necessarily come into play in the passage of the nutritive solution, in the first place into the root, and thence upwards into the leaves.

But before we begin to examine these processes, let us recall to our mind the conditions of Fig. 1. The young root there figured may be looked upon as a compound organ consisting of various parts. First of all we have the actual body of the root, the centre of growth of which lies in the tip near *D*, *P*, and *Pl*. There the cells are in a meristematic condition, like those shown in Fig. 5.

These cells are still very thin-walled and entirely filled with an albuminous substance resembling the white of egg, which is called **protoplasm**. Embedded in this mass we find a denser body, richer in albumen than the general mass of protoplasm, and called the **nucleus** (Fig. 5 *a*). Within the latter we see one or more refringent bodies, the **nucleoli** (Fig. 5 *b*). These delicate meristematic cells, which are constantly dividing, and are thus producing the elements of new growth, become larger and more transparent the farther they are removed from the actual apex. At the point where the first root-hairs are developed, the cells of the region R in Fig. 1 will already have the appearance of the right-hand cell of Fig. 6. Here the viscid mass of protoplasm no longer occupies the entire cell space or **lumen**. It still forms a thick lining (*primordial utricle*) on the inside of the cell-wall, as can be seen in every parenchy-

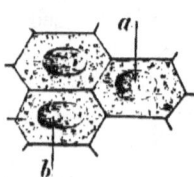

FIG. 5.—YOUNG (MERISTEMATIC) CELLS WITH NUCLEI.

a, nucleus; *b*, nucleolus.

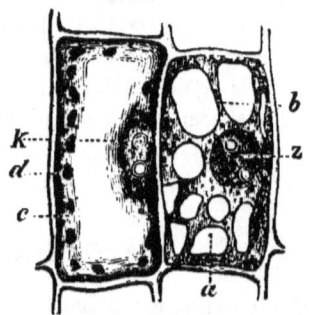

FIG. 6.—YOUNG PARENCHYMATOUS CELLS.

matous cell as long as it continues to live. But in the centre of the viscid protoplasmic mass spaces (*vacuoles*) (Fig. 6 *a*) have made their appearance, and these are filled with a watery cell-sap. As the cell grows, the vacuoles become larger and more numerous, and the mass of protoplasm separating them consequently thinner and thinner, until they present the appearance of delicate threads of protoplasm suspending the nucleus in the centre of the cell (Fig. 6 *z*). Finally, these threads too break across (Fig. 6 *b*), the cell then contains a large central vacuole, and the nucleus with its nucleolus (Fig. 6 *k*) slides towards the cell-wall, which is now evenly lined by the primordial utricle (Fig. 6 *c*).

Cells in this condition, as we find them, for instance, in the older portions of the root cortex, are able to store up other substances, which may now make their appearance. Thus we often find in these cells a certain amount of starch, which makes its way down from the upper portions of the plant, and may be looked upon as a reserve substance, which the plant is able to store up, if the leaves are producing more of it than can at the time be used up. In cells belonging to the upper portions of the plant, which afterwards become green, the green corpuscles (*chlorophyll corpuscles*) make their appearance in the primordial utricle, as may be seen from Fig. 6 *d*.

In consequence of the storage of starch grains within the cells, the cortex of some roots—the carrot, for instance—in the autumn has a snowy white appearance and is full of starch, so that we may look upon the cortex of the root as a storage tissue. This storage is sometimes only of a transitory nature, often limited to a single layer of cells, which surrounds the vascular cylinder. In such a case this layer of cells is termed a starch-sheath, or, if it contains sugar, a sugar-sheath.

The function of the cortical cells as a storage tissue is chiefly seen in that portion of the root where root-hairs have already disappeared and the epidermis has become thick-walled and has lost its absorptive power. Hence we can divide every root functionally into three zones of very unequal lengths. The apex, protected by the root-cap, is the region of **cell-division**, and is concerned in the formation of new tissues. Then we gradually pass into the **absorptive region**, which is also the **region of cell-growth**. Here all the cells increase in size, especially in length, and this region is consequently the chief seat of the elongation of the root. A number of the epidermal cells, too, grow out into root-hairs, which undertake the absorptive function. The absorbed salts and water pass on through the young cortical cells into the vascular cylinder, and through the vessels up into the leaves of the plant. Behind the absorptive region begins the third zone, which reaches back to the insertion of the root, and which comprises the fully developed portion of the root. This older region of the root has now lost its absorptive powers, and the parenchymatous cells of the root serve only as a **storage tissue** for reserve substances, which are of the nature of organic substances, and which have been elaborated in the leaves and have passed down to the root structures.

Of these three zones of the root, the median one, in which the absorption of salts takes place, is undoubtedly the most important for the general economy of the plant. It is represented in Fig. 1 as covered with root-hairs. We must not, however, forget that the production of root-hairs is not essential for absorption, but that the tabular cells of the smooth epidermis are quite capable of absorbing the necessary solutions. But in that case the quantity absorbed by each cell will naturally be very much smaller than that absorbed by a cell which has elongated to form a root-hair, and may increase its absorptive surface a hundredfold.

Every cell of the absorbing epidermis of the root represents a closed space, and yet, as soon as the plant is in want of it, water with its dissolved salts enters through the cell-wall into the closed cell space. This method of passage of gaseous or liquid substances through an uninjured membrane is termed **diffusion**, and it can be explained by the assumption that the cell-wall, just like all other organic structures, is composed of very small particles (*micellæ*), between which there remain very minute interstices, so minute as not to be visible with our strongest optical instruments. The presence of these interstices makes every membrane into a very delicate filter, and allows the passage of all bodies which are small enough to pass through them, and which are attracted by the cell-contents.

The attractive centre of each cell is the substance we have termed protoplasm. The richer a cell is in protoplasm, the greater will be its need for nourishment, and the stronger its power of attraction. As the inner cells of the cortex are richer in protoplasm than the outer ones, they will be stronger centres of attraction, and consequently a current will be set up in the direction of the central cylinder. The vessels of the latter are always completely or partially filled with water, and form a continuous, though sometimes curiously contorted, system of pipes, leading to the leaves, and there they give off to the green cells as much water and salts as the latter require. The leaves lose daily large quantities of water by transpiration, and the vessels which have to supply the leaves must make up the deficiency by absorbing water from below.

We see, therefore, within the vascular system of the whole plant a very active process of suction taking place. This reaches right back into the absorptive regions of the root. There a very active interchange takes place. The parenchymatous cells which adjoin the vessels are able, owing to their rich cell-contents, to absorb large quantities of water, or rather of a solution of nutritive salts. Filled with water in this way, they become turgid, and their **turgidity** causes a pressure to be exerted on the cell-wall. When such a turgid cell borders on a vessel from which water is being withdrawn at one end, as is the case here, it readily forces water into the vessel. But as hundreds of thousands of such cells surround the vessels of the root, and press their superfluous water into the vascular cylinder, a considerable pressure arises, which forces up the columns of water through the vessels into the stem structures. Such a pressure (**root pressure**) actually exists in all roots, and is often of considerable strength. In the case of the vine, for example, a root pressure has been observed strong enough to force a column of mercury to the height of 39 inches.

The root therefore acts continuously like a force-pump. The effect of this pumping action is of course not always so marked as in the case of the vine, in which considerable quantities of water will for days escape from the cut end of a stem (**bleeding of plants**). A strong vine yields during the first day after it has been cut about a quart of sap. If the soil is very damp and the transpiration of the plant is reduced, the root pressure will be increased; the reverse conditions diminish the root-pressure. At times it is so great that it will force drops of water out of the apex or teeth of the leaves. This occurs frequently in the autumn in hothouses which have been cooled down considerably during the night, and also in cool houses in which the heating has not yet been started. At such times drops of water will be found to make their appearance at the tips and on the teeth of leaves of many Aroideæ (*Calla æthiopica*, for instance), of grasses, Fuchsias, Pelargoniums, &c.

The suction due to the transpiration which takes place from the surface of the leaves and the forcing action of the root pressure are compensating factors in the passage of water through the vessels. If the leaves exert but little suction, the root pressure forces the water more strongly up the stem. A given amount of pressure will act more vigorously in this way the smaller the friction is within the vessels. The resistance due to such friction diminishes with the increasing diameter of the vessels. It is for this reason that climbing plants (*Vitis, Tecoma*, &c.), which have to carry large quantities of water to considerable distances, are provided with very wide vessels.

In the case of narrow vessels, however, another force comes into play to a considerable extent, namely, **capillary attraction**. The attraction of water by the wall of very narrow vessels causes the liquid to creep up in the thin tubes, where it is often retained in considerable columns, even though air should enter at the bottom. In such cases we often find small columns of water interrupted by columns of air, the whole forming an interrupted column of water and air, termed **Jamin's chain**.

This chain-like arrangement, which is formed very frequently when rapid transpiration takes place above ground without sufficient supply of water from the root system, makes it possible for small quantities of water to be distributed over the whole length of the vascular system, so that the various organs will all be supplied, should they draw still further on the limited supply. In this way they can be preserved from drying up altogether. This would not be possible if the water were to sink to the bottom of the vessels. Besides this, the small quantities of water in the tubes can move about more easily when forming a Jamin chain, and can therefore more readily supply the thirsting cells at various levels of the stem. For when the outer temperature rises, and consequently the temperature within the plant gradually increases, the air expands within the vessels and raises the intervening drops of water.

Thus, while the sucking and forcing actions of the conducting tissues are specially efficient in the rapid conduction of large quantities of water, the capillarity of the vessels and the formation of Jamin's chains are of especial benefit in the case of a reduced water supply. But besides this, a continuous interchange of water takes place within the walls of the cells and vessels, except in the case of extreme drought. The water which is contained in the interstices of the membrane of one cell or vessel can be partly attracted by the wall of an adjoining vessel, and so the moisture will spread evenly through the walls of these two adjoining elements. When the cell contents can part with no more moisture, the cell-walls can still effect an interchange among themselves. This is true not only of the vessels, but also of all lignified cells.

We cannot conclude the consideration of this part of the subject without referring to the occurrence of special valves which we find in this arrangement of suction and forcing pumps. The valves are represented in this water-conducting tissue by thin portions of the cell-walls, which are usually termed **bordered pits**.

In the cells of all the tissues which make up the vegetable body, we can easily recognise, even if the walls are only slightly thickened, that they are made up of two layers, which are physically and chemically different one from the other. We can recognise an outer thin but hard membrane, which was the first to be formed, and is hence termed the **primary membrane**. Within the latter will be seen a less resistant but thicker layer, which seems to form a padding on the inside, and which is called the **secondary membrane**. This layer, however, is not of the same thickness throughout, but presents smaller and larger areas, at which it seems more or less interrupted. At such points the thickening of the cell-wall, which always takes place when a cell passes from its youthful condition into its mature state, may have been defective, or may not have taken place at all. If these areas have the appearance of small holes or pits in the secondary membrane, the cells or vessels are termed "**pitted**;" if the apertures are numerous and large, so that the secondary membrane has the appearance of a net spread over the inside of the primary membrane, we speak of reticulate elements. In scalariform, annular, or spiral cells or vessels, the secondary membrane presents a ladder-like, annular, or spiral appearance.

We shall get an idea of the method of thickening of the cell-wall from the representation of a thick vessel cut longitudinally in Fig. 7 (*lower g*). The figure represents a longitudinal section through the horizontal runner of the Liquorice.

On the left are the narrow wood fibres tightly wedged together (*b*), and adjoining a few layers of wood parenchyma (*hp*). Then follows a thin pitted vessel (*g*), with more wood

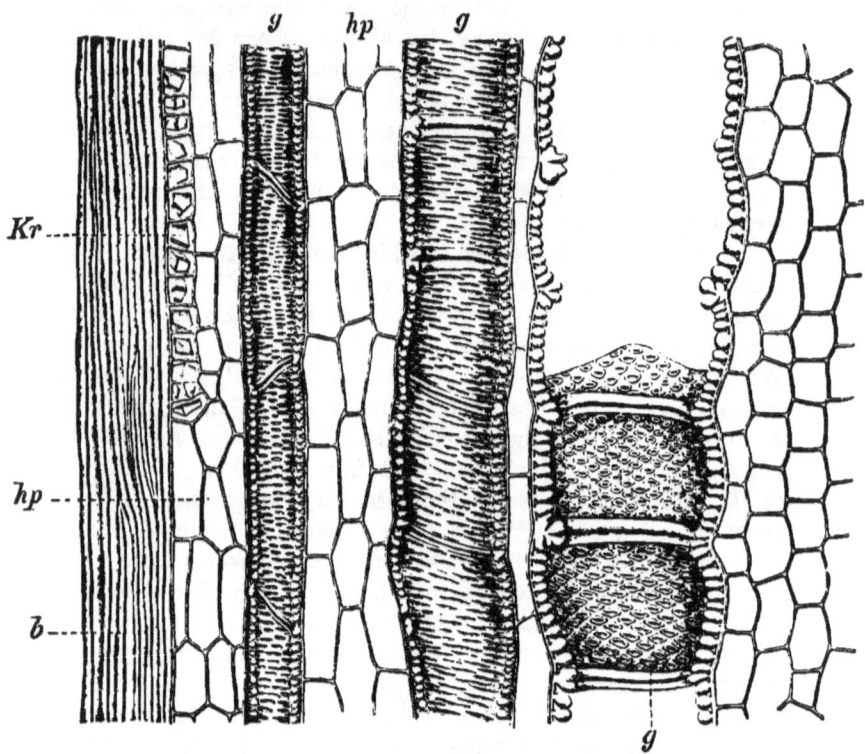

Fig. 7.—Longitudinal Section through a Runner of *Glycyrrhiza glabra* (Liquorice).
g, vessels with slit-like pits; *hp*, wood parenchyma; *b*, wood fibres (*after* Tschirch).

parenchyma to the right of it. Still more to the right is a large closed vessel (*g*), which is seen from the outside. Here the thinner portions of the wall appear as somewhat spirally arranged simple slit-like pits. How this appearance is formed on the wall of the vessel we may gather from a glance at the widest vessel lying farthest to the right (*lower g*), which has been cut longitudinally, so that its interior is visible. First of all we are reminded by the figure of the origin of vessels from superposed cells ; for we can still see the several barrel-shaped segments, the transverse ends of which run round the inside of the vessel. Each such ring consists of two light-coloured rings separated by a dark line. This latter represents the original cell-wall (primary membrane) which separated the two cells, and upon it the lighter-coloured thickenings (secondary membrane) have gradually become deposited. This same layer of thickening has, however, not been so evenly deposited over the remainder of the vessel as it was on the rim of the original transverse walls, but is perforated by small funnel-shaped pits, the structure of which will perhaps be best understood by an examination of the lateral walls, which are the only portion preserved in the upper portion of the vessel. These walls appear to consist of a thin continuous outer wall beset with a number of beads. This bead-like arrangement is exhibited by the section of the secondary membrane which is perforated by the pits, the space between two of these bead-like bodies representing the pit itself, the broader end of which lies towards the primary membrane. We see, therefore, that these pits of the vessel are not actual pores, but only holes in the inner secondary membrane, covered in by the thin outer or primary membrane of the vessel.

This process of thickening, which takes place with the maturing of the cells, is an absolute necessity. For how should such a colossal growth of millions of cells piled one upon the other preserve its shape if the cells remained in their soft and youthful condition ? The whole structure would collapse by its own weight into a shapeless mass.

Hence it becomes necessary to strengthen the separate cells and vessels, and this is done by the formation of the secondary membrane on the inside of the cell-wall. But it is also useful, if not necessary, that this secondary membrane should not be continuous, but should leave pores or pits at intervals, so that the water may pass easily through the thinner portions of the cell-wall, for its passage through the thicker portions of the cell-wall could only be accomplished with great difficulty.

Thus in the construction of all cells two ends have to be kept in view: firstly, the rigidity of the organ, and secondly, the facility of passage of the nutritive fluids.

The structure of the wall, however, shows us other very useful arrangements, and to these belong the valves which have received the name of bordered pits, and which regulate the amount of water which may pass from cell to cell. First of all we have to grasp the fact that in the case of two adjoining cells or vessels the pit in one cell always corresponds in position to a pit of the adjoining cell. This is not shown in Fig. 7, because at the side of the large vessel there is no second vessel, but only thin parenchymatous cells. The current of water has, therefore, only to pass through the thin primary membrane, which is common to the two cells and separates the two pits one from the other. The communication of two pitted vessels or cells resembles, therefore, a room which has been divided into two compartments (cells) by a piece of outstretched calico. But it has been found desirable to strengthen the calico by covering it with a layer of mortar on either side. So as to ensure communication between the two compartments, however, glass rods and glass funnels have been placed at right angles to the sheet, meeting each other, so that the space they protect should not be covered by the lime. When the thickening is completed, we may consider the rods and funnels broken away or withdrawn, and the wall will then be ready for its functions. Wherever a glass rod was previously fixed, a pit will be found, at the bottom of which the sheet will be visible. Two such pits on either side of the sheet will form a cylindrical canal; but where two glass funnels touched the sheet with their broad bases, the canals will become narrower towards the room itself and the two canals will therefore form a lens-shaped hollow, in the centre of which will be stretched a piece of the calico sheet.

If we were to introduce a hand into such a narrow canal, and to push the sheet towards the other room, we should be able, if the sheet were sufficiently extensible, to push our fist so far that it would come in contact with the narrow neck of the funnel-shaped opening on the other side. In that case the fist would close the opening.

But suppose that, in the process of thickening, a small mass of lime had fallen upon this portion of the sheet, and had become hardened in the form of a small disk in the middle of the separating membrane; then if this mass is pushed towards the other side, it, instead of the fist, would close the aperture.

In this somewhat rough illustration, the two halves of the room represent two adjoining cells (Fig. 8), the sheet is the primary membrane (*o*), and the deposit of lime the secondary thickening (*s* and *s'*), the funnel-shaped hollow towards the other cell represents the widening of the bordered pit (*c*), the portion of the sheet within the hollow the **closing membrane** of the pit, and the small disk of lime on the separating membrane stands for the thickening of the closing membrane, usually termed the torus (*t*).

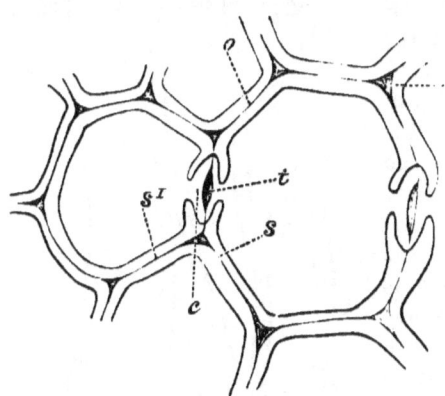

FIG. 8.—TWO TRACHEIDS SEEN IN TRANSVERSE SECTION.

o, primary membrane; *s* and *s'*, secondary membrane; *i*, intercellular space; *t*, thickening on the pit membrane; *c*, cavity of the bordered pit.

Now, if the water is at the same pressure in two adjoining elements of the tracheal system—for instance, in two tracheids—the closing membrane will remain in its original median position, and if there should be only a small excess of pressure in either cell, the water will easily pass through the thin portion of the closing membrane, and the equilibrium will be readily re-established. Such an easy functioning of the valve-like arrangement we may assume to take place in such wood as that of Conifers, which has no true vessels, but in which the wood consists only of wood tracheids with a row of bordered pits (Fig. 9). In this figure we see the pits in surface view, and there the light central ring represents the free portion of the canal (*t* in Fig. 8), while the darker outer ring represents the portion (c) below the overhanging margin of the secondary membrane. This form of cell distinguishes the Conifers from dicotyledonous plants, which have fully developed vessels. In the latter, it often happens that in the summer the leaves withdraw so much water from the vessels that the latter are almost devoid of water and filled with air. If this condition of drought lasts for a considerable time, the delicate membranes will be in danger of drying up, and thus of losing their power of conducting water. But in the case of the funnel-like constriction of the bordered pits, the water is retained in the pit by capillarity, even when the vessel itself is already dry, and consequently the danger of destroying the functioning of the valve is avoided. If afterwards one cell becomes filled with water while the other is still devoid of water, a considerable pressure is exerted by the full cell; this causes the closing membrane to bulge out towards the empty cell. In doing this, the membrane, becoming extended, enlarges the area of passage. If a vessel becomes

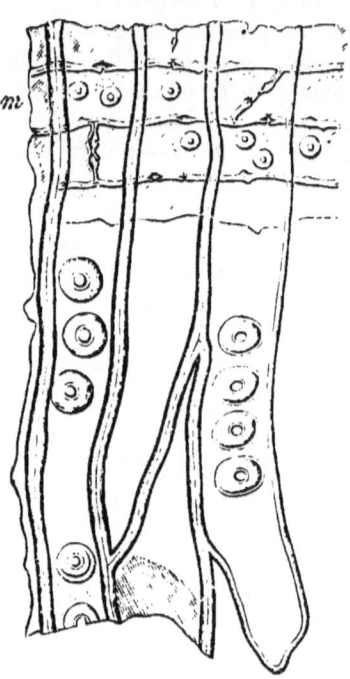

FIG. 9.—PORTION OF A THIN LONGITUDINAL SECTION THROUGH THE WOOD OF THE PINE, SHOWING BORDERED PITS IN SURFACE VIEW.

m, medullary-ray cells, showing the lens-shaped sections of the bordered pits.

filled very suddenly and with great force (as, for instance, if the air in one of the vessels becomes very much rarified), the pressure can become so great that the closing membranes are burst. In such a case the small thickening in the centre lessens the danger, for as soon as the membrane is considerably pushed in, the thickening presses against the wall of the funnel, and the pressure is spent on the thickened portion, which will resist any pressure likely to occur under such conditions.

This useful system of valves is only developed between two water-conducting elements; if, however, a vessel adjoins a cell containing protoplasm or reserve substance, the bordered pit will only be developed on one side. If the parenchymatous cells are permanently rich in water, while the vessel is poor in water, or even filled with air, then the high pressure (turgidity) of the parenchymatous cells will cause the cell-wall to bulge out through the pit into the lumen of the vessel, and the latter will become filled by protrusions on all sides from the adjoining cells. These become divided off from the original cells which gave rise to them by transverse walls, and enlarge so as to completely plug up the vessel. This filling of the vessels with bladder-like cell protrusions (thylloses) takes place especially when the wood system becomes damaged, and it affords a protection of the injured portion against the effects of injurious atmospheric conditions.

Let us now sum up shortly the results of this paragraph. The vessels and vessel-like cells, all tracheal elements therefore, are at the commencement filled with water, but can be partially and temporarily emptied by the transpiration from the leafy portion of the plant. In that case they become filled again by the pressure of water from the parenchymatous cells surrounding them. If the vessels become emptied of water, the air within them must take its place, and fill the cell or vessel. In doing this it becomes greatly rarified; and if it be cut under water, it will suck up the latter with great avidity (*negative pressure*). Besides this rising of the water by suction, an upward forcing takes place by root pressure. Lastly, the capillarity of the vessels, and the changes effected in the Jamin's chains which they contain, are of considerable importance. These two physical forces make a continuous flow of water in the vessels possible, if the leaf structures only use up a moderate amount of water compared with the amount absorbed. They also enable a passage of water to take place when the vascular system is considerably emptied by an inordinate amount of transpiration. The regulation of pressure in a lateral direction of the columns of water of adjoining vessels is effected by the valve-like action of the bordered pits, and the slow movement of water through the insterstices of the cell membrane.

These functions are only performed by the cells and vessels of the wood (*xylem*) contained in the central cylinder of the root. The bundles, however, possess also a soft bark (*phloem*), which is concerned in the passage of the organic substances. We shall discuss the method of functioning of this portion of the bundle in dealing with the structure of the stem, and shall now pass on to the consideration of the substances which enter the plant dissolved in the water taken up by the roots.

CHAPTER III

THE NUTRITION OF THE ROOT

§ 4. What substances must be present in the soil for the continuous and effective nutrition of plants?

In order to find out what substances are necessary for the growth of cultivated plants, a number of chemical analyses have for years been made of the most widely differing plants, and it has been found that only a very small number of chemical elements are present in all of them. From this small number, too, a few must be excepted, which occur in some individuals grown in special localities, and which are therefore, as we might say, accidentally taken up by some plants, while equally robust specimens of the same species are devoid of them. The results of these analyses caused physiologists to try to cultivate plants in media the composition of which was known, but which in themselves had no influence on the plant. The choice lay between distilled water or quartz sand which had been sterilised by heat. To these media were added the substances which had always been found in the ash of plants. These substances were added in various forms, and one or other was occasionally omitted, so as to find out which substances must always be present in order that the plant may grow successfully. These water and sand cultures resulted in showing that even some of the substances which were present in the ash of all cultivated plants might be absent without lessening the growth of plants; and hence we may assume that they are only present in all our cultivated plants because they are present in every soil, and are therefore taken up by all roots.

The knowledge thus obtained enables us to divide the substances found in the ash of plants into three categories:

(*a.*) the absolutely necessary or **essential nutritive substances**, which must always be present in sufficient quantity and in soluble form: these are oxygen, hydrogen, nitrogen, carbon, sulphur, phosphorus, potassium, calcium, magnesium, iron, and probably also chlorine; (*b.*) **subsidiary substances**, which are present in every ash, but are not indispensable for the growth of plants: sodium, silica; (*c.*) **admixtures** which are due to the nature of the soil: such as zinc, copper, nickel, aluminium, cobalt, manganese, lithium, strontium, and barium.

§ 5. What is the effect of the various nutritive substances on the plant, and in what form do they enter it?

All tissues of the plant contain **oxygen, hydrogen, and carbon.** The first two substances are not only necessary to the plant in the construction of its tissues, but they are essential for purposes of conduction; for they make up the water which pervades the whole plant. The percentage of water present in the various parts of the plant is naturally very variable. Succulent herbaceous tissues possess 60 to 75 per cent. of water, while the fungi often possess only 5 to 8 per cent. of organic substance, the rest of their weight being made up of water. As most of the nutritive salts are contained in the soil, we can easily form an opinion of the importance of water, which is the medium by which the roots have to absorb the salts. These solutions pass through the vessels of the plant, and the adjoining cells can absorb water from the vessels, though they have often to pass it on to neighbouring cells, and thus a constant absorption of the water contained in the soil is brought about. Nearly all the water which passes through the plant, carrying with it the nutritive salts, is taken up by the root. This water is ultimately given off as water-vapour, or is chemically decomposed and used up in the construction of new material. Only in cases of extreme dryness of the soil are the leaves able to make use of the heavy deposits of dew.

In the case of **carbon**, however, which must be looked upon as the chief constituent of the vegetable tissues, forming as it does one half of their total dry weight, the soil is of no use to the plant as a source of this element. The atmosphere, however, which surrounds the plant affords it all the carbon which it requires. The leaves absorb carbon from the air in the form of carbonic acid, and decompose the latter in their cells under the influence of light, giving off the surplus oxygen.

Nitrogen is the substance which is chiefly concerned in the building up of young tissues, but it is also present throughout the whole life of the cell in the substance which we termed protoplasm.

Alkaloids, which are present in many plants, and asparagin, which is concerned in the regeneration of the albuminous substances, are formed from the protoplasm, and contain nitrogen. Most of our cultivated plants have to take their nitrogen from the soil; it is only the leguminous plants which are able to take up the nitrogen through their leaves from the atmosphere, of which nitrogen is the chief constituent. As far as our present scientific knowledge goes, the leguminous plants can subsist on the nitrogen which they take out of the air; while cereals, fruit-trees, and indeed all other Phanerogams, must obtain this substance in some soluble form from the soil. Nitrates seem to be the most suitable form of salt from which plants obtain their nitrogen. Ammonia, which can probably be absorbed in minute quantities even in a gaseous state, is less suitable.

Sulphur and **phosphorus** are of some importance in the formation of albuminous substances; both are taken entirely from the soil, where they are found in the form of calcium salts. This is especially the case with sulphur, which is widely distributed as calcium sulphate (gypsum), while calcium phosphate is of less frequent occurrence, and being only very slightly soluble, it is advantageously replaced by a potassium salt. In some places phosphoric acid is found very largely in the soil in combination with iron. Plants which are deficient in phosphorus generally assume a red coloration.

Potassium, which is chiefly found in very young tissues, and disappears in the older ones, is probably connected with the formation of the **carbo-hydrates**, *i.e.*, with those organic substances which consist of carbon, hydrogen, and oxygen, and in which the last two elements occur in the same proportions as they do in water.

Starch, sugar, and cellulose are such carbo-hydrates which are transmutable within the plant. If a plant does not find any potassium in the soil, its growth ceases, and the leaves do not develop the power of forming starch within the green colouring bodies, the chlorophyll grains. A plant which has perhaps shown no growth for months, owing to the absence of potassium, will recover in a few days after that element has been supplied to it; even after a very few hours starch will begin to make its appearance in the chlorophyll grains. The very soluble salts of this important nutritive substance are liable to become washed out of the soil if watering is too continuous.

Calcium, the foundation of all combinations of lime, is chiefly concerned in the strengthening of the cell-wall, within which it is often deposited in visible crystals of carbonate of lime. But besides this, it is important in fixing the oxalic acid which is produced by plants, and which is poisonous to them. In this process of fixing, crystals of calcium oxalate are formed, which in many plants are stored in continuous rows in the cells of the cortex immediately surrounding the channels through which the plastic material passes, and showing therefore their intimate connection with the metabolic processes. For as the organic substances passing downward from the leaves through the cortex undergo manifold changes, chiefly of the nature of further oxidation, it is probable that the last function of the lime is to fix the oxalic acid, which is formed by the oxidation of the carbo-hydrates, and thus to render it harmless. Calcium is taken up by the plant in the form of sulphates, phosphates, and carbonates. In the case of water-cultures, calcium nitrate may be advantageously used.

Magnesium, which of all the nutritive substances is most nearly allied to calcium, but cannot replace it in the vegetable economy, seems to have quite a different effect on the plant. It seems to work together with the nitrogen in the formation of protoplasm, and to influence the formation of chlorophyll; for plants cultivated without magnesium only form yellowish-

C

green chlorophyll granules, and do not form new cells very readily. In the case of decomposition of succulent portions of plants, very often only those parts which are very rich in nitrogen will be found to contain crystals of magnesium phosphate. Magnesium, just like calcium, enters the plant in the form of sulphates, phosphates, and other salts, especially in combination with chlorine.

Iron is necessary in the building up of chlorophyll. As it is the function of chlorophyll to form new plastic material under the influence of the sunlight, it is natural that the absence of iron, which is shown by the paleness of the leaves, should cause a cessation of assimilation. But as iron is very widely distributed in the soil, such paleness (*etiolation*) is of much less frequent occurrence than might be supposed. In diseases of all kinds yellow foliage may make its appearance, and we must not be surprised if in many cases the addition of soluble iron salts remains without remedial effect.

To the subsidiary substances we must also reckon **chlorine**, which is not looked upon as a nutritive substance by many physiologists; still many observations tend to show that this substance is essential to the healthy growth of the plant. Water-cultures devoid of chlorine compounds do not thrive, and in some cases direct indications of disease make their appearance. From the present data, we can only assume that the presence of chlorine retards the absorption and ultimate deposition in the tissues of calcium, but that, on the other hand, it accelerates the passage of phosphates, and thus increases the supply of food material at the growing points.

If much phosphoric acid is used up, the nitrogen and the soluble carbo-hydrates are more rapidly utilised by the protoplasm for the formation of albuminous substances. Hence we find that a supply of chlorine salts (calcium potassium and magnesium chlorides) causes long and succulent shoots to be produced, which, however, are poor in starch.

§ 6. **What part is played by the subsidiary nutritive substances and the occasional admixtures to the nutritive salts ?**

Sodium must be looked upon as a subsidiary food substance, though it is absent only in very few plants, and is taken up by some groups of plants (shore plants) in enormous quantities in the form of sodium chloride (common salt), and is found in their tissues together with potassium. But in spite of these two elements being closely allied, they cannot in any way replace one another. Thus, if a plant is without potassium, a supply of sodium will be of no use to it. The roots take up the very soluble sodium salts in great quantities if they require the acids with which the sodium is combined, and store up the sodium in another form as a waste product. But we gather that this substance is of little importance in plant economy from the fact that even those plants which always present a large amount of sodium in their ash, such as the saline plants of the shores, can be cultivated to perfect development in a soil free from sodium.

Silicium, which is present in all soils in the oxidised form of **silica,** behaves just like sodium. Silica is generally found in the soil in the form of crystalline (insoluble) substance, but it occurs also in an amorphous soluble form, and can in this condition be easily absorbed by the roots. Being so widely distributed, it is found very generally in plants, and, like lime, and often together with it, it is used to strengthen the cell-walls. Such cell-walls are often so strongly impregnated with silica, that in some cases (*Equisetum*) these plants may be used for polishing purposes. The stems (*culms*) of our cereals and the leaves of sedges are provided with epidermal cells so stronly protected with silica that a hand passed over them will be severely cut. The enormous quantity of silica contained in grasses and sedges was formerly adduced as a proof of the necessity of silica for the strengthening of the cell-walls. But now that it has been possible to cultivate cereals in solutions devoid of silica just as well as in the open field, this inference breaks down. The laying of cereals, which was formerly attributed to a deficiency of silica in the lower joints or nodes of the culm, is known now to be due to the incomplete thickening of the cell-walls, in consequence of deficient illumination of the base of the stalk.

Of the substances which are occasionally present in plants, and which are due to the habitat of the plants, **iodine** and **bromine** must be mentioned as always present in marine plants. The sea-water contains only very small quantities of these substances, while more than 1 per cent. of the dry weight of some sea-weeds consists of these substances (poisonous for other plants) combined with potassium. Some chemists have found both these substances in bog and also in land plants, while others have not been able to trace them in other specimens of the same species. This shows how dependent the appearance of these substances is upon the nature of the locality in which the plants grow.

Boron has been observed in traces in the Grass-wrack (*Zostera*) and in some sea-weeds, while **fluorine** has been found in the Club-moss (*Lycopodium clavatum*), in the Horse-tail (*Equisetum*), and in grasses.

Of the lighter metals which are usually found associated with potassium, **lithium** and **rubidium** (not cæsium) have been found in some plants, the former, for example, in the tobacco-plant and in thistles; the latter in turnips and in the leaves of tea and coffee. But in spite of their affinity to potassium, they can in no way replace the same, but are, in fact, poisonous if they occur in any appreciable quantity.

Allied to calcium is **barium**, which has been recognised in the ash of several trees and shrubs, and **strontium**, which has been discovered in several sea-weeds. The former of these two substances is harmless to many plants in fairly large quantities; the latter acts as a poison even in very small quantities. In spite of the wide distribution of clay in soil and in rocks, its chief constituent, **aluminium**, is confined in its occurrence to very few plants (lichens and club-moss). This element occurs both in rocks and also in the product of their breaking up, in the soil, almost exclusively in the insoluble form of a double silicate of potassium and aluminium.

Of the metals which are widely distributed in the vegetable kingdom, we must mention first of all copper, which has been found in the wood of various trees (oak, beech, lime, &c.), and in some herbaceous plants. Not so widely distributed, but occurring in higher percentages in the ash of plants, we find zinc, which changes the entire appearance of some plants (*e.g.*, *Forma calaminaria* of *Viola lutea*), but only acts detrimentally when occurring in larger quantities. The leaves of plants which suffer from zinc-poisoning are discoloured by small rusty-looking spots, which after a while spread over the entire surface of the leaf. This metal generally enters the plant as a soluble sulphate. Lead is less harmful to plants, but arsenic is poisonous in very small quantities, decreasing very rapidly the absorptive power (*osmotic activity*) of the roots.

From the above account, which is by no means exhaustive, it will be seen what a number of elements can be present in plants. Some of these substances are stored in large quantities without injuring the plants in any way, while others prove harmful when present in very much smaller quantities. From this fact we may gather that roots are not able to choose their nourishment; if, therefore, poisonous substances reach the roots in a soluble form and can penetrate through their cell-walls, the plant will gradually be poisoned. But even if the substances are harmless in themselves, they will always have some effect on the general economy of the plant; sometimes this influence will become apparent externally. Thus we have already seen that zinc plants possess a different appearance— in some cases a differently coloured flower. The same may be said of plants growing in salty regions (*halophytes*); they are, it is true, able to flourish without absorbing any salt, but they only develop their succulent character when they have absorbed some sodium chloride. Similar experiments with regard to silica have been made on cereals, and they tend to prove an indirect influence of this substance on the general growth. The grains were better developed and the other food substances were more completely used up.

We must, therefore, not look upon the substances only occasionally absorbed by the roots as entirely useless for the general nutrition of our cultivated plants. Their influence, sometimes accelerating, sometimes retarding, will depend upon the nature of the plant, and vary at different periods; for one species will be unaffected by a large amount of a certain substance, the same amount of which would be injurious to another species or stop its growth entirely. If, therefore, we hear of plants the ash of which is shown by chemical analysis to contain a strikingly large amount of such a substance, we must not suppose that this species needs so much of that substance for its actual growth, but all that we can logically conclude is, that this species can absorb without harm to itself the stated amount of that substance. Plants rich in salt or saltpetre, in silica or zinc, are therefore not plants which need these substances in such large quantities, but which have the power of resisting the over-loading of their tissues with these waste products. The same may be said of the plants rich in lime.

If, therefore, a plot of land, the soil of which has a special character owing to the preponderance of one of these substances, is to be planted afresh, those plants should have the preference which can endure the largest amount of the substance in question. What is said here of the several elements of the soil is equally applicable to the other properties of the soil (the facility of absorbing heat, the power of retaining water, &c.). The constant occurrence of certain kinds of plants in localities characterised by the preponderance of certain substances is explained by the greater power some plants have to adapt themselves to the given chemical and physical conditions of the soil.

For practical purposes the occurrence of lime-, silica-, saltpetre-, or salt-loving plants should be noted by horticulturists and agriculturists, as it affords a clue to the condition of the soil. Such soils, characterised by the preponderance of one substance, have the disadvantage that a number of cultivated plants do not thrive on them on account of the excessive richness in this one substance. If, therefore, a crop should fail in such soil, we know where, in the first place, to look for the cause.

§ 7. In what form does the root find the nutritive substances in the soil?

Every soil, representing as it does the product of disintegration of various rocks, consists of a mass of fragments of various kinds, endowed with certain physical powers of attraction. A mechanical analysis of the soil enables us to separate any soil under cultivation into its coarser constituents, the skeleton or **framework,** and finer **earth,** which can be washed out of it. According to the size of the fragments making up the framework, we can distinguish the coarsest constituents as coarse gravel; then follow gravel of medium coarseness, and fine gravel or coarse sand. The fine sand would belong to the earthy portion, which shows the characteristic physical properties that are of importance for the growth of plants, such as the power of absorbing and conducting heat, capillary attraction, and the power to attract and retain gases and liquids. It is the varying proportions in which the various constituents of the framework are mixed with the earthy portion of the soil which gives to every soil its characteristic structure. The **structure of the soil** is important to the plant in the first place, because it regulates the access of air to the roots, which require air just as much as the upper portions of the plant. Some plants, however, require a large amount of air circulating among their roots, while others are content with a much smaller amount of air, because they need only a very small quantity of the really active constituent of the air, namely, the oxygen. A loose soil, in which the several particles lie loosely packed side by side, will not only stimulate the activity of the roots more than a binding soil, in which the particles are closely fixed together, but it will also favour the processes of decay and decomposition. The most **favourable structure of a soil** for cultivation is one which is not too loose to retain the water which is necessary for growth, and yet is not so close as to have its interstices filled with water under normal climatic conditions, which would prevent the access of the necessary amount of air to the roots.

Of the nutritive substances, the one of chief importance for the roots is the water, which circulates in the interstices of the soil, and is there retained.

The capacity of a soil to retain water without allowing it to escape in the form of drops (the retentiveness of the soil) increases with increasing fineness of the particles; but the attraction which the particles of the soil have one for another can be overcome even in fine-grained soils with great water capacity; large quantities of water are forced into the soil; the particles then become separated, and the soil becomes soft and pasty. In the case of clays, less often of loamy soils, a complete dissolution of the soil may take place. This may be very serious in the case of seed-beds, as the soil between the young plants is washed away, and the roots are laid more or less bare. Such a condition of the soil will also cause the sowing of crops to be postponed, and the yield of the harvest in our country often depends upon the time of sowing. For if plants have to produce a good stock or a number of large leaves before they can yield a good crop, a retardation of the time of sowing may prevent the proper development of such growth, and an early dry summer will stimulate the development of flowers, which will be scanty and poor. In hot and dry summers we see the peas which have been sown late attacked by mildew and smut, the kolrabi becoming woody, and the lettuces running to seed.

The heterogeneous particles of the soil possess in different degrees the power of attracting by physical and chemical means, and of retaining the substances which are washed into the soil. This can be demonstrated by a very simple experiment. If a drain-pipe is filled with clay, the highly-coloured liquid manure from a dunghill will be entirely discoloured by passing through it. The organic compounds are retained by physical forces of attraction, while inorganic compounds are chemically fixed. This power of the soil to bind the various organic and inorganic salts is termed its **power of absorption**.

The root of the plant will not alone absorb the water contained between the particles of the soil, but can also make use of the salts which have been absorbed by it.

(a.) *Potassium.*

Of the nutritive substances which have to be added to the soil, potassium is most readily absorbed by the soil; the next in order are ammonia, magnesium, sodium, and calcium. Of the acids, the very essential phosphoric acid is very energetically retained. This is not so much the case with carbonic acid, and still less with nitric acid. Sulphuric acid and hydrochloric acid are not fixed at all by the soil, which will, however, absorb silicic acid.

The absorptive power of the soil depends very much on the quantity of basic hydrated silicates which are soluble in hydrochloric acid (zeolithic). They contain the necessary lime, sodium, magnesium, and potassium, but are almost useless to the roots on account of their insolubility. They are, however, decomposed by the carbonic acid contained in the soil or by compounds of nitric or sulphuric acid with alkalies or alkaline earths, and the resulting carbonates, sulphates, and chlorides are readily soluble, and can therefore be easily absorbed. During the decomposition of certain silicates hydrated iron oxide and hydrated aluminium oxide are formed, and these compounds as well as the hydrated silicates enable the soil to separate the potassium from its most stable combination with hydrochloric, nitric, and sulphuric acid and to retain it. These iron and aluminium compounds absorb very little potassium, however, without the presence of such silicates, but the soil absorbs it very actively if aluminium phosphates is present. Potassium is, however, scarcely absorbed at all by the humic acid compounds, by carbonate of lime and of magnesium.

The potassium salts which are absorbed by the soil are able to form combinations with other nutritive substances, and therefore to fix important substances in the soil. Thus **kainit** (a sulphate and chlorate of magnesium) can be used to fix nitrogen. The actual potassium salt (potassium sulphate) which takes part in this process plays a quite subsidiary part. As soon as ammonium carbonate begins to form in the soil, the acid of the magnesium sulphate combines with the ammonia; the ammonium sulphate so formed is still very soluble. But if the application of kainit was preceded by the addition to the soil of a manure containing phosphates, the more insoluble ammoniacal magnesium phosphate is formed. In stables the giving off of ammonia is often mitigated by strewing gypsum, the sulphuric acid of which combines with the ammonia to form the ammonium sulphate. The same effect may be more speedily and more completely attained by the addition of kainit; and a similar action to that of the magnesium sulphate of the kainit would be produced by sulphate of iron (green vitriol).

In light soils, therefore, in which there is little humus, clay, or other absorptive substances, manuring with kainit would enrich the soil by causing the ammonia, which becomes volatilised in the decomposing manure, to be fixed and retained in the soil. The common salt which is also present acts as a solvent and distributing agent for other salts.

(b.) *Phosphoric Acid.*

This very important nutritive substance, which counterbalances the excessive addition of nitrates, is immediately absorbed in the soil, if in solution, by the heavy metals and the earths, as these substances will at once combine to form insoluble phosphates. The acid is then generally found combined in the form of calcium and magnesium phosphates, and in smaller quantities as iron and aluminium phosphates. The richer, therefore, a soil is in the first place in carbonates of lime and magnesium, the more we shall be able to saturate it with phosphoric acid. And the same effect will be produced by those silicates (sesquisilicates) which yield on decomposing hydrated aluminium oxide and iron oxide. If lime is present in the soil in the form of gypsum, it can take up phosphoric acid, gradually giving off sulphuric acid as it does so. The phosphoric acid is only available to the roots directly in the form of its soluble combinations with alkalies (potassium and sodium).

(c.) *Nitrogen.*

This substance, as important as potassium to plant life, represents 79 per cent. of the air we breathe, the remaining

21 per cent. consisting mainly of oxygen. This **uncombined nitrogen** is only available in any appreciable quantity for leguminous plants, which possess, as we now know, the faculty of growing in a soil entirely devoid of nitrogen; for though it may possibly take part in the nutrition of other cultivated plants, at least to a very slight extent, it must be absorbed by their roots in a combined form. On the other hand, the small green Algæ, which are present in every soil, seem to have the power of making use of and of elaborating the free nitrogen, and may therefore be looked upon as nitrogen purveyors to the soil. The nitrogen compounds which are most important for the nutrition of plants are ammonia, nitrous oxide, and nitric acid. This combined nitrogen, however, which alone can be assimilated by the roots, is not derived from the atmosphere except to a very slight extent. The decomposition of nitrogenous organic substances liberates ammonia into the air, and the lightning flashes of a thunderstorm cause some free nitrogen to combine with ozone to form nitrous acid (ozone and hydrogen peroxide also being produced). Besides this, ammonium nitrite is formed in every process of oxidation, and even during the evaporation of water.

When the water-vapour of the air condenses into drops, ammonia is enclosed in them, and more is absorbed by the drops when falling as rain through the air. Dew, hail, and to a lesser extent snow, all contain ammonia. Rain, too, especially during a thunderstorm, contains some nitric acid. These available nitrogen compounds are thus washed into the soil by natural agencies, or may be (but only to a very slight extent) absorbed directly out of the air. But the absorptive power of a soil for nitrogen depends upon its composition. The several constituents of the soil do not only vary in their capacity of absorbing nitrogenous compounds, but they also vary according to the form in which they occur. Humus compounds, gypsum, carbonates of lime, and magnesia, for instance, take up very little nitrogen; but if humic acid combined with calcium occurs in the soil together with decomposing silicates, a very considerable absorption of nitrogenous compounds takes place. From experiments (of Bretschneider) it has been calculated that if water absorbs about 0.64 lbs. of ammonia per acre, gypsum will absorb only 0.23 lbs., carbonate of lime 2.62 lbs., pure quartz 0.93 lbs., but iron oxide 10 lbs. If, however, pure quartz is mixed with a brown humus compound (*ulmin*), it will absorb, by the addition of only 1 per cent. of ulmin, 14 lbs.; by the addition of 5 per cent., 92 lbs. of ammonia. This demonstrates the indirect **influence of humus compounds.**

Although in artificial cultures in distilled water or sterilised sand, plants thrive exceedingly well when supplied with inorganic food substances only, and a proof has therefore been given that these plants can grow without any decomposing organic substances, still the figures just quoted indicate that the organic substances of humus are of considerable value in the nutritive processes. In certain cases humus is absolutely necessary. To the plants which live entirely on organic food matter belong all the fungi, which take their nourishment either from living organisms (*parasites*) or from decaying organic matter (*saprophytes*). Besides these cryptogamic or flowerless plants, many more highly organised plants afford us examples of plants which must necessarily be supplied with organic food matter. The mistletoe illustrates the parasitic habit on a living host plant, while decaying organic food matter is essential for the nutrition of the humus-growing plants which have either very little chlorophyll in their tissues or are completely devoid of it. Examples of these are the Yellow Bird's-nest (*Monotropa Hypopitys*), and the yellowish-brown Orchids (*Neottia nidus avis, Epipogum aphyllum*, and *Corallorhiza innata*). In connection with the mistletoe we must not omit to mention those green plants, rich in chlorophyll, which are able to lead an independent existence, growing and flowering in the soil, but which can on occasion further their nutrition by absorbing the sap of other plants. To these belong the well-known inhabitants of moorland meadows, such as the Eyebright (*Euphrasia*), the Yellow Rattle (*Rhinanthus*), and the Cow-wheat (*Melampyrum*). The roots of these plants, if they come into contact with other roots, develop small suckers, which abstract substances from the roots of the host plant. The desire to supplement the normal nutrition by a parasitic absorption of organic nitrogenous substances and carbo-hydrates is therefore apparent from these instances.

After the discovery of these facts, the next question to be solved was whether any of our cultivated plants which do not absolutely require humus for their growth, might not be benefited by it. This problem led to some very interesting culture experiments, which consisted in watering one set of plants with an extract of humus, while another lot of the same plants received the ash of the same quantity of humus. Now, if only the mineral constituents of the humus were concerned in the nutrition of the plants, both series of cultures should have yielded the same result. But this was not the case. The plants which had received the humus extract yielded a crop twice as large as those which received the salts only.

This experiment, therefore, must be taken as a proof that plants in the open reap a distinct advantage if they receive their food substances in part at least in an organic form, and we have a scientific confirmation of a practical and popular saying, that "The dung put into the ground accounts for nine-tenths of the farmer's pound."

Of course it is not ascertained as yet whether the greater vigour of growth is due to the absorption and the more rapid assimilation of the organic compounds, or whether it is due to the fact that combination of the humus compounds with inorganic substances causes them to absorb more of the very important ammonium salts. At all events, we must hold that the presence of humus in the soil greatly stimulates the growth of the plants, and that the view held up to the present with regard to the importance of the humus compounds must be considerably extended. According to this older view of Grandeau's, the function of the constantly changing humus compounds was to keep a sufficient quantity of nutritive substances in solution in the soil and make them available for the roots. Chemically these humus compounds are still very insufficiently determined.

But in considering the efficacy of humus, we must not only take into account the substances from which it is formed, but we must know at what stage of decomposition it has arrived.

This fact can be illustrated by experiments in which plants are treated in one case with humus in its raw or fresh condition, and in the other with humus which had been subjected to the action of steam (212° Far.) for several hours. The same quantity of soil after the action of the heat yields a crop many times in excess of the former.

The mechanical effect of the humus upon the structure of the soil we shall deal with in speaking of stable manures, but to the general consideration of the efficacy of humus we must append an account of the curious modification of the roots of many of the humus-loving plants. This special form of root is termed **mycorhiza**, and it differs from the normal root structure by the fact that the lateral rootlets, instead of being long and slender, are thick and short, more branched and interlocked, and form a nest-like structure. These rootlets, too, are usually devoid of root-hairs, so essential for absorption; but their entire surface is covered by a dense felted mass of fungal threads, so that the root itself does not seem to be directly in contact with the soil at all, but can only take up its nutriment by means of the fungal threads. As the root grows in length at its apex, the pseudo-parenchymatous threads of the fungus increase in number too, so that the slightly developed root-cap is constantly covered in by a dense cap of fungal hyphæ. This fungus being devoid of chlorophyll cannot, as has been proved, absorb or utilise any raw or inorganic salts, and probably derives all its nourishment from the organic compounds of the humus. If, therefore, we do not like to assume that this fungal covering becomes mechanically saturated with the inorganic nutriment of the soil and passes this on to the cells of the root, we will be forced to the conclusion that a large number of cultivated plants actually derive their nourishment from the organic substances of the humus, and do this either by preference or by necessity.

How widely distributed in nature this arrangement of mycorhiza is may be gathered from the fact that all our forest trees, such as oaks, beeches, birches, alders, Conifers, and also poplars, willows, and limes, are provided, where they grow in humus, with mycorhiza. They will, however, develop normal rootlets and root-hairs if they are grown in solution or in purely mineral soil, or if the humus has been sterilised by heating, and the fungus is therefore killed. According to A. B. Frank, upon whose extensive investigations these remarks are based, mycorhiza is also present in heaths, in most orchids, in many Liliaceæ and Smilaceæ, and in a large number of Compositæ, Labiatæ, Primulaceæ, Umbelliferæ, Rosaceæ, Leguminosæ, and Ranunculaceæ.

The fungi which form the mycorhiza belong to those kinds which usually occur in the humus of woods and fields. They do not seem to act in any way injuriously on the root, but, on the contrary, stimulate it to renewed growth, and no doubt the root in its turn stimulates the growth of the fungus, nourishing it with its own sap. We have here, therefore, an example of the living together or partnership of two organisms (*symbiosis*), each of which benefits the other. The fungus, too, is not always restricted to the outside of the epidermis, or to making its way between the epidermal cells (*ectotrophic mycorhiza*), but it may live exclusively within the living cells of the root (*endotrophic mycorhiza*). In this latter case the roots need not necessarily present the short, thick, and much-branched condition, but may be long and very slender, and devoid of cortical tissues. The fungus fills the epidermal cells, which then lie immediately outside the central vascular cylinder. In those roots which possess cortical parenchyma, the endotrophic mycorhiza occupies a ring-like zone of cells, from which fungal threads run towards the outer surface of the root.

While the symbiotic association of these green plants with a fungus is only formed when they grow in humus, and is more largely developed as the soil becomes richer in humus, it occurs always in plants devoid of chlorophyll, and there the mycorhiza is either ectotrophic or endotrophic. In these plants, therefore, the mycorhiza appears as a necessary means of absorption, while in the chlorophyll-containing plants it is only occasional and additional means of nutrition. Thus we find in the absorption and assimilation of decomposed organic matter a parallel to the conditions obtaining in parasitism, *i.e.*, in the absorption of living organic matter. We have obligatory humus absorbers (*Monotropa*), to which plants humus is indispensable, and obligatory parasites (mistletoe), contrasted with occasional absorbers of humus (forest trees) and occasional parasites (cow-wheat, yellow rattle). As a similar and equally important special method of nutrition in the vegetable kingdom, we must, in our present state of knowledge, consider the **root tubercles of leguminous plants.**

If we take up carefully peas, lupins, beans, robinias or true acacias, and other Leguminosæ, which have been grown in sand, and rinse the sand from their roots, we find the main root in some cases, in others the lateral rootlets, bearing curious fleshy tubercles, which are sometimes of characteristic shape for different plants. These generally have the appearance of a rudimentary tubercular rootlet, and in the first stages of their development appear to the naked eye very much like a lateral root breaking through the cortical tissue. But in reality their origin is a very different one. These tubercles arise by active cell division of the inner layers of the cortex, while the lateral roots arise from the pericambium. The pericambial layer, however, only takes part later on in the formation of the tubercles. At the periphery of the tubercles there will be observed a layer of cambiform cells, which produce towards the outside large thick-walled cortical cells, and towards the inside a smaller-celled tissue, and also a ring of fibro-vascular bundles, which join on to the bundles of the root. As the tubercles have for some time a meristematic apex, by which they continue to grow, their similarity to a short succulent rootlet is still further increased. But, of course, they are not provided with a root cap. In the older tissues behind the meristematic apex there gradually accumulate large quantities of oval or rod-shaped, undivided or forked bodies, which are capable of moving about in water for many days. These small bodies can be shown to consist of albuminous substances, and therefore characterise the tubercles as storage tissues for nitrogenous material (albumens), which are gradually used up by the plant during the ripening of the seeds.

The tubercles appear in greater numbers the poorer the soil is in humus and soluble nitrogenous substances, especially nitrates. If the soil is sterilised by heat and soluble nitrates are added to it, leguminous plants will be able to grow, but will not produce tubercles.

The most important fact, however, is that leguminous plants exhibit a very healthy development and yield a good crop in a soil which contains much too little nitrogen for such a crop, or even when the soil is entirely devoid of nitrogen. In plants grown under these conditions the production of root tubercles is very great. It follows from this that these leguminous plants must have absorbed the nitrogen necessary for their development from the air, and that this power of absorption of free nitrogen has some connection with the presence of the root tubercles. We know, therefore, of some plants which are able to make use of the air as an inexhaustible source of nitrogen, and can absorb from it sufficient free uncombined nitrogen to provide for an entire and rich crop.

This power of making use of the atmospheric nitrogen is, as far as experiments have so far shown, not possessed by any other cultivated plants; they do not seem to derive any appreciable advantage from it, and for their normal development the presence of a large amount of nitrogen in the soil is essential. The richness of their crop is approximately proportionate to the amount of nitrates in the soil. But there is in reality no fundamental difference in the methods of nutrition between leguminous and other plants; for the former are able to nourish themselves entirely from nitrates if they are plentifully present in the soil, but they possess in the root tubercles a special provision by which they can supply themselves with nitrogenous food when other plants would not be able to do so.

This view is the outcome of extensive and conscientious experiments of Hellriegel, upon which experiments we base the subsequent remarks.

With regard to the faculty of the Leguminosae to live upon the atmospheric nitrogen, it has been proved that the plants showed a complete and normal development if they were grown uncovered in the open in a soil with little or no nitrogen compounds, but in which the other food substances were present. Their development was furthered, too, in a soil devoid of nitrogen by the addition to the soil of an extract in distilled water to the extent of 1–2 per cent. of a soil in which the same leguminous plants had formerly been grown. After such an addition, the development of the plants was not only normal, but often indeed luxurious; but their mode of development was very different from plants of the same species and origin which absorbed their nitrogen in the form of nitrates from a sterilised soil. For while the latter exhibited a continuous and even growth from the moment of germination, the plants grown in a soil devoid of nitrates passed through a very marked and characteristic phase, exhibiting the signs of starvation, and this was followed, after a longer or shorter period, by a very active growth.

Such a beneficial extract cannot, however, be made from every soil, and even an active extract becomes powerless if it is boiled, or only heated up to 70° C. There is a difference, too, in the way in which leguminous plants of the same species react to extracts from different soils. Thus an extract of a soil in which peas and clover had been regularly grown for a considerable period, but on which lupins and Seradella had never yet been grown, was only beneficial for the growth of peas in the experimental soil, but was perfectly useless for the growth of lupins or Seradella. Lastly, the very interesting experiment was tried of growing a leguminous plant in such a way that one half of the root grew in a solution devoid of nitrogen, but supplied with such a soil extract, while the other half of the root grew in a nutritive solution of the same composition to which the soil extract had been added after it had been sterilised by boiling. The result was that root tubercles were only developed in the solution which had not been sterilised.

All these results point to the conclusion that the development of root tubercles—that is, the formation of centres of assimilation of atmospheric nitrogen—depends on the development of organisms within the roots of leguminous plants. These organisms are probably different for the different kinds of Leguminosæ, and multiply in the soil in which these leguminous plants have once been grown. This supposition is strengthened by the fact that for years various observers have described within the substance of the root tubercles small bodies which have been looked upon as parasitic organisms. In the last few years, too, successful inoculations have been made; these consisted in introducing the point of a needle into one of the root tubercles and then pricking with the same needle the healthy root of a plant of the same species. Even microscopically the spreading of the organism from the point of inoculation has been observed; at least in the case of the lupin the tubercle-producing parasite is known to form swarm-spores, which pass over into a zoogloea stage, and this organism is now known to science as *Rhizobium leguminosarum* (*Bacillus radicicola*, Beyerinck).

In consequence of Hellriegel's experiments, it is probable that several kinds of Rhizobium occur among the Leguminosæ. The curious circumstance that the plants, when grown in a soil devoid of nitrogen, pass through a starving condition, in which they draw upon the reserves of the most important assimilating organs before they form the root tubercles, is perhaps also capable of explanation. The roots of all plants contain nitrates. (The upper portions of plants only contain these substances occasionally, chiefly in the case of annuals.) If we could, therefore, prove that the organisms which produce the tubercles did not develop in the presence of nitrates, the following would probably be the state of the case. If the soil is well manured, the parasites enter upon a passive stage (zoogloea stage); the plants grow without drawing upon the atmospheric nitrogen. If the soil is poor in nitrogen, the parasites become partially active, and the formation of tubercles commences. Should the soil be entirely devoid of nitrogen, the roots will only contain nitrates as long as those contained within the seed last out. The moment the root becomes poor in nitrogen (that is, the time when the plant draws upon its own organs), it affords the best culture medium for the parasites, and these produce large numbers of tubercles which assimilate the atmospheric nitrogen.

Horticulturists may, therefore, in the case of pot-culture of leguminous plants, dispense with any sort of nitrogenous manuring.

§ 8. Why and how should we replace the chief nutritive substance in the soil?

(*a*) *Fallow.*

The harvests which are taken from the ground in cultivating the soil consist either of fruits, seeds, or inflorescences, in which case the remainder of the plant is returned to the soil, or else the whole plant may be withdrawn from the soil for economic purposes. No soil of ours can endure for any length of time the loss of mineral substances, which form part of the crop, without being replenished, for no soil contains sufficient substances to cover the loss. The soil will therefore become impoverished more or less rapidly, according to the nature and requirements of the various crops. A certain number of substances, such as hydrogen, oxygen, and carbon, are replaced by the natural wetting and aëration of the soil. Of the mineral substances, lime and iron can be renewed for the new crops by the natural processes of decomposition taking place in the soil; but potassium, phosphoric acid, and nitrogen, which are taken up in large quantities from the soil into the younger organs (as in vegetables), or into the ripened storage tissues (fruits and seeds), must be replaced in some way or other.

In former times, when agricultural methods were less exacting and exhausting, the method employed to restore the productive activity of exhausted soil was to let the soil lie fallow, the natural means which the soil possesses to recoup its losses by rest. Under the same climatic conditions, fallow land is moister and warmer than cultivated soil; but a greater moisture and higher temperature increase and accelerate the processes of decomposition which are going on in the soil, which is indicated by the fact that fallow land is richer in carbonic acid than land in cultivation. Now the more rapid decomposition of organic substances on the one hand, and on the other hand the greater amount of carbonic acid thus formed, which causes a more rapid solution of the mineral constituents of the soil, increase the amount of soluble substances by which the next crop will be benefited. As fallow land has also a smaller covering of plants, it does not lose so much of the natural condensations by evaporation from the surface of the plants; it will therefore also contain, when it begins to be cultivated again, a larger amount of water, *i.e.*, the means of transport of the soluble substances. Lastly, a certain amount of enriching in nitrogen takes place through the wild leguminous plants.

This beneficial effect of lying fallow is, however, only noticeable in rich soils; a sandy soil may, indeed, become impoverished by lying fallow. For the soluble organic substances resulting from decomposition are not readily absorbed by the sand, and may be washed down by strong rains into the deeper layers of the soil, where they will only be accessible to a few deep-rooting plants. Under such conditions, therefore, the soil must always have a dense covering of plants. The latter, however, will of course use up both water and nutritive substances, and store up nothing for a future crop. A heavy soil, too, may be impoverished under abnormal climatic conditions by lying fallow. If, for instance, a continuous and excessive rainfall has closed up the already very small interstices of a clay soil for a considerable period, the process of decomposition of organic substances may become altered, and substances are formed which act injuriously on the root structures. We shall return to this subject in dealing with the watering of pot-plants.

Lying fallow, therefore, only has the desired result on a heavy soil, and under normal or dry climatic conditions; and the present methods of cultivation have caused this means of enriching the soil to be entirely given up, and now recourse is had almost exclusively to the direct addition of nutritive substances by **manuring**.

(*b.*) *Inorganic Manures.*

Originally stable manure was the only food material directly added to the soil; now-a-days, however, artificial manures have come more to the front. These do not, it is true, contain, like stable manure, all the food substances of plants, but they furnish the soil with all the most essential and most important ones in a more concentrated form. In practice it will be best to add, in the first place, all the three most important food substances, and then to judge from the nature of the crops which manures should in future be given in increased, and which in lesser quantities.

According to the poorness of the soil, potassium and phosphates should be used to the extent of 25 to 40 lbs. per acre. The nitrogen in the form of soluble nitrates should be added in half the above amount.

If you notice a tendency in the plants to run too much to leaf, and to remain green for too long a time, decrease the amount of potassium and nitrogen, or increase the amount of phosphoric acid.

The chief forms of mineral manures are **ammonium sulphate** and **Chili saltpetre** (sodium nitrate). The former, to be in condition for absorption by the plant, must be transformed in the soil into a nitrate, and should therefore be in the soil some considerable time before the sowing takes place. The Chili saltpetre, which is readily soluble, but is scarcely at all absorbed by the soil, and is therefore liable to be washed into the deeper layers by heavy rain, must be placed on the ground very shortly before the sowing or planting of the field. As it is immediately effective, it may be recommended for occasions where plants which have suffered from external disturbances (cold, hail, or drought) are just beginning to recover, and require some stimulant to accelerate their growth.

Potassium is commercially obtainable in form of **raw potassium sulphate**, or as a purer and more concentrated salt. In whatever form it is used, it must be mixed sufficiently deeply in the soil, not placed on the top, and must be introduced into the soil at a time when a long damp period is certain to follow (autumn or winter). If the soil is not rich in lime, it is advisable, when manuring with the raw potassium sulphate (which contains a large amount of magnesium chloride) to add a considerable amount of powdered quicklime.

Which of the commercial potassium salts should be used in any special case will depend partly on the distance of transport. If the distance between the market and the estate is great, the more concentrated forms are of course more advisable.

Phosphoric acid, too, can be obtained in various forms.

The so-called raw phosphates and bone-meal contain this nutritive substance in a very insoluble form, and must therefore be mixed with the soil a considerable time before sowing. They are only to be recommended for fields known to be sufficiently damp, and for autumn sowing, where the chief productivity takes place in the spring. If bone-meal is to act more rapidly, it should be left to ferment in heaps before use. The quickest effect is obtained from **superphosphates**. If it is desirable to add nitrogen as well as phosphoric acid to the soil, a mixture of ammonium sulphate and superphosphates, so-called ammoniacal superphosphate, may be used. The third chief nutritive substance, which contains potassium as well as nitrogen and phosphoric acid in readily soluble form, is **Peru guano**. These last two substances, on account of their being immediately available by the plants, are used as top-dressing for weak crops, and by gardeners for wholesale growing of herbaceous plants, especially foliage plants. From the first of these two preparations we must, however, distinguish the raw ammonia superphosphate, which must only be employed in fields which are well aërated and have a considerable moisture, for this preparation contains a considerable amount of rhodan-ammonium, which acts injuriously on plants.

A preparation which has more recently been largely used is the Thomas-slag, which is derived from ironworks, and which must be ground as fine as possible. It contains about 40 to 60 per cent. of phosphate of lime in different combinations (tricalcium phosphate, &c.), besides silica, clay, iron oxides, and traces of magnesium, chlorine, and sulphur. The lime, besides occurring in combination with phosphorus, is present in considerable quantities combined with silica, and also in the form of quicklime. This manure is also most effective in a damp soil, and when added at an early period. In a dry soil it is only rendered soluble after a considerable time.

(c.) *Organic Manures.*

In spite of the evident beneficial results of inorganic manures, and in spite of the fact that they are indispensable in our present system of agriculture, they cannot replace stable manures entirely. Our agricultural efforts must aim at getting as much animal dung as possible, and to use the mineral manures as additions to the former. For besides taking into account the chemical constitution of the soil, we must also bear in mind its physical or mechanical structure, and the latter is greatly improved by stable manure, but is rarely benefited — sometimes indeed deteriorates — through mineral manures. As an example of the deterioration of good soil under such conditions, the continuous manuring with saltpetre and common salt might be cited. After successive manurings of a clay soil with saltpetre, the finest particles of clay will be found washed together and deposited in a solid and compact form.

As the nitrates are so easily washed out of the soil, the latter often becomes so caked that plants cannot be grown in it. Manuring with saltpetre may, therefore, at first produce fine green crops, but soon a sudden deterioration will be noticed. Similar results follow on the manuring with common salt, which is carried on in the case of certain crops. Soils which contain basic carbonates in large quantities (alkaline soils) have a consistency which renders their working quite impossible.

(d.) Stable Manures.

As it is the first endeavour of an agriculturist to produce the most favourable physical conditions of a field and to preserve them, because otherwise chemical changes take place which are detrimental to the life of plants, the chief point which has to be considered in manuring a field is to find a manure which will serve the above purpose best, and this is certainly **stable manure**.

The ideal condition of a soil is one in which it resembles a sponge, and in which it will retain the greatest amount of nutritive substances and water, without losing its capacity of absorbing air.

Heavy soils can be made to approximate this condition by addition of straw-containing manure, while a light soil requires the shorter decayed manure. Of course such manuring is not always seasonable, and in each case the gardener or farmer has to consider whether he ought to change the mechanical nature of his soil, or need only enrich his field with concentrated mineral manures. For though stable manure contains all the necessary ingredients, it does not contain them in so concentrated a form as artificial manures. To give an approximate idea of the nutritive value of stable manures, we append the mean values published in Mayer's "Agricultural Chemistry," and based upon Heiden's calculations. The following amounts are contained in the excrements of the

HORSE.

	In the Dung. Per Cent.	In the Urine. Per Cent.	In a Mixture of both. Per Cent.
Water	76	87–92	76–79
Organic substance	21	6.9	19
Nitrogen	0.4–0.5	1.5	0.6
Potassium	0.35	1.6	?
Phosphoric acid	0.32	.0	?
Total ash	3.15	3.1	3.15

OXEN.

	In the Dung.	In the Urine.	In a Mixture of both.
Water	82–85	92–95	86–89
Organic substance	14.6	3.2	10–12
Nitrogen	0.17–0.38	0.3–0.9	0.34–0.44
Potassium	0.05	1.3	...
Phosphoric acid	0.15	.0	...
Total ash	1.9	3.1	2.1–2.4

SHEEP.

	In the Dung.	In the Urine.	In a Mixture of both.
Water	57–75	85–90	67.0
Organic substance	24–37	5–10	27.5
Nitrogen	0.5–0.7	1.3–2.5	0.9
Potassium	0.1–0.4	2.1–3.3	?
Phosphoric acid	0.3–0.6	Trace	?
Total ash	3.0–5.7	3.2–6	5.4

PIG.

	In the Dung.	In the Urine.	In a Mixture of both.
Water	77–84	98	82
Organic substance	10–15	1	14
Nitrogen	0.7	0.23	0.6
Potassium	0.27	0.7	?
Phosphoric acid	0.4	?	?
Ash	6.5–7.5	1.0	3.7

	Pigeon Dung. Per Cent.	Fowl Dung. Per Cent.	Duck Dung. Per Cent.	Goose Dung. Per Cent.
Water	62	65	53	82
Organic substance	31–32	21–26	40	14
Nitrogen	1.2–2.4	0.7–1.9	0.8	0.6
Basic salts	2.0–2.2	1.2–4.0 (0.9 Potassium)	0.4	3.1
Phosphates	3.0–4.2	2.0–5.0 (1.2 Phosphoric acid)	3.5	0.9
Ash	6.0–7.0	9.0–14.0	7.0	4.0

	Fresh Human Excrements. Per Cent.	Closet Manure. Per Cent.
Water	92.9	97.0
Organic substance	5.7	1.5
Nitrogen	1.06	0.35
Potassium	0.22	0.20
Phosphoric acid	0.23	0.28
Ash	1.37	1.5

If we compare with these values the most perfect of artificial manures of organic origin, guano, we find that it contains—

	Peru Guano. Per Cent.	Baker Guano. Per Cent.	Sombrero Guano. Per Cent.
Water	14.8	3.3–10.5	2.9–10.1
Organic substance and ammonium salts	52.4	6.6–9.5	4.7–6.6
Nitrogen	14.4	0.3–1.0	?
Phosphoric acid	13.5	37.0–40.3	32.5–39.6
Alkaline salts—			
Potass	7.4	0.1–0.6	Traces
Lime	...	39.1–43.5	22.7–51.7
Ash	32.8	80.8–89.7	85.0–90.0

Of the pure mineral manures, which only contain a few of the nutritive substances, the percentage can be gathered from their advertisements.

In stable manures, however, the animal excrements do not form the chief mass, but the latter is formed by the substance which serves as medium of absorption or litter. The latter generally consists of straw, so we append an analysis of straw and of other substances which are used for the same purpose, such as pea-stalks, dry leaves, and fresh leaves of pine or fir : [1]—

---

[1] It cannot be considered the object of this book to fully discuss all the materials used as manures, although the practical importance of that subject is fully recognised. But we are here only concerned in illustrating the nature

	Corn-Stalks, Per Cent.	Pea-Stalks, Per Cent.	Autumn Leaves, Per Cent.	Fresh Leaves of Fir or Pine, Per Cent.
Water	12-21	12 17	13-15	47.5
Organic substance	75-83	82-83	78-81	52
Woody fibres	29-53	34-52	11-16	?
Nitrogen	0.3-0.9	2.0	0.8-1.4	0.5-0.9
Potassium	0.5-1.1	1.1	0.15-0.4	0.03-0.1
Phosphoric acid	0.2-0.3	0.4	0.2-0.3	0.1-0.2
Ash	3.0-8.0	3.0 4.0	4.2-5.7	0.8

In the use of stable manure, which we will now consider as a mixture of animal excrements and litter, our chief object is to use it to its best advantage. This is only possible if we allow the constantly occurring processes of decomposition to run their proper course. This they will do if a sufficient amount of oxygen is present. Then certain processes of oxidation are set up, which we term decomposition, or which, in the case of nitrogenous substance, we sometimes describe as decaying. If, however, the supply of oxygen is deficient or ceases entirely, other processes will take place, and the products of these are often injurious to plant life.

The above processes of decomposition, which take place by the agency of bacteria, break up the organic substances into carbonic acid, water, ammonia, and free nitrogen, which substances are partly given off as gases, leaving the mineral constituents in an easily assimilated form. The ammonia is transformed by bacteria into nitric acid; the formation of carbonic acid takes place even in the absence of oxygen, but is more plentiful up to a certain degree in the presence of oxygen.

Light and heat are, of course, not without their effect. But while light acts in such a way as to retard the formation of nitric acid, probably because the formative bacteria are averse to light, the increase of temperature up to $37°$ C. causes an increased amount of nitrification (formation of nitric acid) to take place. The maximum evolution of carbonic acid takes place about $60°$ C.

Moisture also furthers the decomposition of organic sub-

of manuring by the substances most widely used, and this only in so far as it is of importance in the understanding of the physiological processes of the plant. The same is true of other portions of the subject which belong more especially to the chemistry of agriculture.

stance, as long as the soil remains porous enough to allow the access of oxygen; but still in a thoroughly saturated soil the formation of carbonic acid does not cease entirely.

Of course the absolute amount of carbonic acid and nitric acid produced depends mainly upon the nature of the substance which is being decomposed and upon the substances with which it may be mixed. The rapidity of the decomposition of various substances used as manures can be roughly gathered from the following figures, which represent the volumes of carbonic acid contained in 1000 volumes of air after 1 gramme of the substance has been decomposing for twenty-four hours: —Steamed bone-meal liberates 31; Peru guano, 24; Pigeon-droppings, 26; fresh swine-dung, 14; partially decomposed horse-dung, 12; fresh cow-dung, 13, and after three months about 9 volumes of carbonic acid; while peat gives off only 2, sawdust 5, ground horn 6, dead leaves 7–8, pine leaves 9, corn-stalks 17–19, and pea and bean stalks 22 volumes of carbonic acid. Though, generally speaking, the addition of quicklime during the later stages of decomposition becomes advantageous on account of the combination of the lime with humic acid, which decomposes more rapidly than the humic acid alone, quicklime generally retards the initial stages of decomposition (Wollny). In the same way, mineral acids even in small quantities retard decomposition, while weak solutions of alkaline carbonates stimulate decomposition. The latter will also be stimulated by small additions of sodium nitrate, but is retarded by large amounts of this salt. Common salt (sodium chloride) always retards decomposition.

As a general guide for the rapidity of decomposition, we may place the manures in the following order. The most rapid of decomposition are bone-meal, flesh-guano, flesh-meal, the droppings of birds; straw and stable manure are slower; still slower are leather scrapings, horn, dried leaves and sawdust; peat is slowest of all to decompose.

Other things being equal, the heat developed will be proportional to the rapidity of decomposition. Horse-dung heats the soil most effectively, cow-dung least. Decomposing cornstalks give off less heat than the stalks of the pea, which contain more nitrogen.

§ 9. How can the soil best meet the requirements of the roots for air?

It is not only because the soil contains an extraordinarily large quantity of the nutritive solutions in its pores that we have compared the ideal soil with a sponge, but also on account of its permeability for air. The roots must breathe, and therefore the soil on which our crops grow, must be of such a nature that down to a certain depth there must be a sufficient interchange between the air contained in the soil and the outer atmosphere. This explains the advantages of the addition of organic humus-forming substances to all soils, whether they be light or heavy. The nature of the air contained in the soil is not, however, immaterial. If the processes of decomposition in the soil take a wrong course, gases detrimental to the root may be formed; and as we are able, whenever we work or manure the soil, to change its structure and its porosity, we have every reason to pay attention to the interchange of gases.

The air contained in the soil has not the same composition as the atmospheric air; for within the soil oxygen is continually being used up by the processes of decomposition of organic substance, and the roots themselves constantly absorb some and give off other gases. Besides these two causes, various gases are physically fixed (absorbed) by the small particles of the soil. Of these gases, we are at present only interested in two—oxygen and carbonic acid. Nitrogen, and its combination with hydrogen (ammonia), we have previously dealt with.

We may assume that, generally speaking, the air of the soil is richer in carbonic acid and poorer in oxygen than the atmosphere. The rapidity of exchange depends on the atmospheric pressure, and upon the moisture of the soil. The greater the water capacity of the soil, the greater are the variations in its permeability. If much water is present, the permeability for air is greatly reduced. A loosening of the soil will increase the permeability to a greater extent, the finer the particles of the soil are.

A passing decrease of the permeability is caused by the frost, which causes the water contained in the interstices of the soil to expand and thus decreases the pores. The water itself, in such a case, looses its mobility. If at this time coarse organic substance is introduced, so as to enlarge the interstices, we at the same time introduce a source of carbonic acid, which lasts as long as oxygen has access to keep up the decomposition. A purely mineral soil contains air of about the same composition as that of the atmospheric air. But even if the oxygen of the air is excluded, decomposition still takes place, partly at the cost of the existing organic substance itself, or with the aid of oxygen gained from the reduction of inorganic substances. This process, however, is comparatively unproductive.

Although the oxygen contained in the soil is of chief importance for the aëration of the roots, the carbonic acid present in the air and water of the soil is chiefly concerned in the chemical changes, and therefore indirectly in the physical constitution.

Distilled water, of course, can only take from the soil the soluble substances, and give them off again when it evaporates. The carbonic acid, however, contained in the water enables it to dissolve substances which are insoluble in pure water —as, for instance, the carbonates of lime and of magnesium, ferrous carbonate, and manganous carbonate. The simple carbonate of lime is transformed into bicarbonate; the phosphates of lime and magnesium become soluble in presence of carbonic acid, and the same is the case with phosphates and silicates of iron, ferrous oxide, and the silicates of potassium, sodium, lime, and magnesium.

The carbonic acid of the air, in conjunction with the oxygen, acts in the same way as the carbonic acid of the water. It enables the oxygen to transform ferrous and manganous oxides into the higher oxides, and the resulting ferric and manganic oxides, as they take up more space than the lower oxides, help to break up the mineral substances of the soil. The yellow or brownish iron pyrites (FeS_2) is changed by the oxygen, in presence of water, into ferrous sulphate and free sulphuric acid. When these products become dissolved in the water, they change the carbonate of lime into gypsum, sodium chloride into the sulphate, and the magnesium contained in the insoluble iron and magnesium compound (dolomite) is transformed into the very soluble magnesium sulphate (Epsom salts). In like manner the insoluble tribasic phosphate of lime is converted into the soluble phosphate, and clay is split up into a sulphate and a silicate of aluminium and magnesium.

By this breaking up of substances the number of soluble substances necessary for the active functioning of the roots is greatly increased. To cause them to function most actively, it is necessary to maintain as even a supply of oxygen as possible, and this can only be done by keeping the physical structure of the soil as constant as possible. This is actually brought about by the breaking up of the soil. The hardest mineral cannot withstand a continued action of the carbonic acid. We find large blocks of granite sticking out of the ground, the surface of which seems eaten away, and when touched they often break to pieces. The eroded surface is caused by the action of the carbonic acid on one of the constituents of the granite, the feldspar, from which it extracts in the first place the potassium and the sodium. The feldspar is thus transformed into clay, from which the more resistant pieces of quartz and mica project. The water containing carbonic acid now separates from the silicates the monoxides (ferrous oxide, lime, magnesium, potassium, and sodium oxides), and by uniting with them it liberates some hydrated alumina, which does not readily combine with carbonic acid, and also a small quantity of silica. This potassium containing feldspar (orthoclas, $3 SiO^2(K_2O + Al_2O_3)$) thus furnishes a large amount of the necessary potassium. The rain then washes away the abovementioned monoxides, and also the sesqui-oxides (ferric oxide and alumina), until only a soft finely-divided mass remains, which forms the almost indestructible framework of the soil, consisting of fine granules of quartz sand and silicates.

To give some idea of the permeability of the soil for air, which is constantly changing as the constitution of the soil is being changed by the breaking down of its original constituents, we append a table drawn up by Ammon.

This table shows how many litres of air pass in one hour through a depth of 50 cm. of soil under a pressure of 40 mm. of water—

		Litres.
Clay	⎫	1.62
Caolin	⎬ in powdered condition ⎧ 2.84
Powdered lime mixed with humus	⎭	3.32
Chalk		3.78
Pure broken limestone	⎫ size of particles up to	4.24
Peat	⎬ 0.25 mm.	5.04
Quartz sand	⎭	16.80
Broken-up clay	⎫ particles of 0.25–0.50	30.9
Quartz sand	⎭ mm.	41.04
Quartz sand	⎫ size of particles	92.24
Broken-up clay	⎭ 0.5–1.0 mm.	123.75
Quartz sand	⎫ size of particles	287.57
Broken-up clay	⎭ 1–2 mm.	420.16

This table is interesting in showing the effect of breaking up a soil which is not readily permeable. The broken-up clay is shown to be more porous than the coarse-grained sand; but a small amount of powdered clay added to the quartz sand will greatly reduce the porosity of the latter. In fact, upon the thickness of a layer of clay which lies between two layers of sand depends the porosity of the whole soil, as all the absorbed air must pass through it. The moisture of the air and of the soil, too, control the amount of air which passes through the soil in a given time. Dry air moves more rapidly through a moist soil than air containing a large amount of water vapour, and so air of a given nature passes more rapidly through a dry than through a wet soil; most rapidly, however, through a slightly moistened soil. This latter phenomenon is probably due to the fact that small particles of sand are broken up by being moistened.

How the various physical properties of a soil are changed by the varying proportions of clay and sand in a field, and how important the presence of stable manure is in this connection, will be seen from the following figures of Masure:—

	Garden Soil.	Sand.	Powdered Lime.	Clay.	Stable Manure.
Saturation for rain-water of the various components of the soil, reduced to 1 vol.	0.39	0.30	0.44	0.68	0.56
The saturated soil loses per diem water to the amount of (in millimetres)	4.20	3.70	3.50	4.30	4.50
The hygroscopic capacity of the soil: 100 grammes of soil condense water (amount in grammes)	5.60	2.10	3.60	7.0	41.0
Capacity of soil to absorb heat, expressed by difference of temperature of dry soil *above* temperature of surrounding air (degrees in centigrade)	14.2	10.7	9.0	11.50	14.70
Cooling at night, difference of soil *below* that of air	-1.60	-1.7	-1.8	-1.8	-0.8
Permeability for air, expressed by the amount of oxygen absorbed by 100 grammes of soil (amount in grammes)	14-18	1.6	10.10	15.3	20.3

§ 10. How can we improve our fields so as to obtain the best possible crops?

In practice, fields may be grouped, according to their constitution, into those with a sandy soil, those with a lime soil, and those with a clay soil. Let us examine them in turn.

The characteristic of a sandy soil is that it can be saturated with a relatively small amount of water. The smaller the grains of sand, the less porous is the soil, and the greater is its power of retaining in its lower layers a small supply of water and of condensing the atmospheric moisture at its surface. Powdery sand is harmful on account of its density (imperviousness). Its great power of absorbing heat in its upper layers from the direct rays of the sun is of advantage only in cold situations, usually, however, harmful. Its power of absorbing oxygen is small. To improve fields of this kind, a large amount of stable manure is necessary, as the rain-water is then retained in large quantities, and at the same time the requirements of the plants in this direction are reduced. For experiments have proved that in a well-manured soil the roots need less water for the production of a given amount of vegetable substance than in a soil poor in food substances though rich in water. The roots must take up a certain fixed amount of nutritive matter for every gramme of dry substance which the leaves have to form. If the solution which the root finds in the soil is very dilute, then a comparatively large amount of this weak solution must be taken into the plant, and much water must therefore be passed out by transpiration through the leaves; and, as a matter of fact, a square inch of leaf surface of a plant growing in dilute nutritive solution gives off a great deal more water within a given space of time than does a plant grown in a richer solution. After sowing it is very advantageous to roll a dry and warm sandy soil, as its power of retaining water is greatly increased by this procedure.

The chalky and marly soils are the chief forms of lime-containing soils. The chalky soil consists of sandy lime mixed with a varying amount of powdery lime. From the table contained in the preceding paragraph we see that chalk can absorb more water than sand, and does not lose it so rapidly by evaporation. It is not so readily heated by the sun as sand, and is only slightly more hygroscopic. Chalk soils must, therefore, be treated very much like sandy soils, but they are able to derive more benefit from manuring, and do so more rapidly than sandy soils.

Marls contain an intimate mixture of powdery lime and clay; the properties of the latter predominate and give to the soil its characteristic features. It must, therefore, be treated like a clay soil. It is more fertile than a clay soil, and this is due to the presence of the lime, to which we shall refer again in speaking of the addition of lime to the soil.

Every field may be said to consist of a clay soil if it contains sufficient clay, for the properties of this substance to predominate. This may take place when it forms only 30 per cent. of the soil. It is characterised by its extraordinary power of retaining water, so that it very rarely drys up. If the atmosphere is moist, its great hygroscopic powers prevent its becoming too dry. If a long and hot drought sets in, the ground cracks, and then of course more water is lost. It is as rapidly heated by the sun as is sand, and has the power of absorbing a large amount of oxygen. We see, therefore, that the clay soil has properties which would ensure a rapid decomposition of manure, a process which is, however, retarded by the chief fault of this soil, its great wetness. To counteract this, the soil must be well drained, besides which, deep ploughing and repeated harrowing, to cause an even breaking up of the soil, are very beneficial. A large addition of straw-containing manure is very paying, as it keeps up the porosity and permeability of the soil.

The chief means of improving the soil is, therefore, as we have seen, the use of stable manure, the nutritive properties of which we have already discussed. Besides these, it is proved to be useful on account of its faculty of retaining water. It can absorb its own, and sometimes even more than its own, weight of water. Being very porous, it is true that it loses this water very rapidly, but its great hygroscopic capacity causes it to absorb the necessary amount of this indispensable substance. Dry stable manure can absorb from the atmosphere and retain 40 per cent. of its weight of water, and this absorption of water vapour takes place in greatest quantity towards the morning, and thus from the absorption of dew the soil obtains carbonic acid, ammonia, and nitrates, which are of considerable benefit to the vegetation. Besides its own nutritive value, stable manure has, as we see, the power of further enriching the soil from the atmosphere, and this fertilising power is enhanced by the fact that stable-manure has a greater power than any other substance of absorbing the heat of the sun and the oxygen of the air. It accelerates, therefore, the processes of decomposition taking place in the soil, and thus brings about and keeps up the conditions necessary for the active life of the roots.

Manuring with stable manure must therefore remain the best method of improving the soil for every sort of cultivation, while artificial manures may form useful additions.

Where market-gardening is carried on on a very large scale, so that occasionally there may be a lack of stable manure, it will be found useful to plough in some crop before it has matured its seeds. Leguminous plants, as, for instance, lupins, are most useful for this purpose, as their power of forming nitrogenous subtances from the nitrogen of the air will further enrich a poor soil. Spurrey and buckwheat are often used, or on good soil white mustard may be grown for this purpose.

Very often a field may be greatly improved by the addition of lime, marl, or gypsum.

Heavy and caking clay soils which are poor in humus often stick to the plough and other implements when wet, and when dry form large impracticable clods. Such a soil may be improved by ploughing it up in the autumn and allowing it to lie in its rough condition during the winter; but the more friable condition which it then assumes is soon lost during the summer after heavy rains. If the addition of stable manure, humus, or peaty soil is impossible, then lime or marls should be added.

The loosening of the soil which is thus brought about is probably due to the chemical action of the lime. If it has to be employed rapidly during a dry season, so that the natural breaking up of the limestone cannot be awaited, and if no powdered form is available, the following procedure can be recommended :—The burnt limestone should be filled into baskets and these plunged into water until no further air bubbles appear (3 to 4 minutes). Then it should be piled in heaps. The pieces break up of themselves, and the limestone (slaked), which had given off its carbonic acid when heated, now forms a white powder of calcium hydroxide (CaH_2O_2) or so-called slacked lime, which is soluble in 730 parts of cold, or 1300 parts of boiling water (lime-water).

The lime attacks the silicates of the clay, decomposes them and liberates soluble potassium salts. It also accelerates the decomposition of organic substances. This is the reason why a field treated in this way uses up its organic manure more rapidly than a field to which no lime is added. A lime soil may therefore be said to be of a devouring nature.

With regard to the manipulation of the lime, it should be strewn very evenly over the ground, either by hand or with a spade, and this operation should not be carried on during windy weather. It will be found advantageous to lay the lime on the stubbles in the autumn and then to plough it in. If forced to wait until the spring, it is best to strew the lime over the fields a considerable time before sowing, as soon, indeed, as the soil has become slightly dry. Small quantities (5 to 10 cwt. per acre) added after every five years are more beneficial than a single heavy addition of lime, as the decomposition of humus becomes so active that very little is left for subsequent harvests. Of course the above-mentioned quantity must only be looked upon as an approximate value; the actual quantity will depend greatly on the nature of the soil. Great care is needed in treating light sandy soil with lime, while on a tenacious clay soil large quantities may be used. The most conspicuous and rapid results of this treatment will be shown by a soil rich in humus but poor in lime upon which the Sheep Sorrel (*Rumex Acetosella*) grows abundantly. In such soil the lime will act directly as food material.

As in the process of marling the lime contained in the marl is also the active principle, it is evident that a soil rich in humus and in clay will stand it much better than a light sandy soil. To the latter a larger amount of clay marl than of calcareous marl or shell marl may be added. The latter two forms of marl are more advantageous for clay soils. The much-dreaded over-marling will only occur when too small an amount of stable manure is added; we must remember that this substance is the basis of the permanent fertility of the soil.

Besides the increased decomposition of organic substances by the lime, which results in an enormous increase in the production of carbonic acid, the lime fixes many free acids which are injurious to vegetable life, transforms the ferrous into ferric oxide, and causes a greater absorption of basic food substances. The bases are contained in the soil in the form of hydrated silicates and humus salts. The bases, if they are to combine with the humus compounds, must be united to carbonic acid. Now the lime promotes the formation of such carbonates. Besides these indirect effects of marling, there are some direct actions which must be mentioned. A direct enriching of the soil is effected by the potassium, phosphoric acid, and magnesia which every kind of marl contains.

The addition of gypsum, which was already practised by the Greeks and Romans, is effective in making the potassium of the soil available to plants as a sulphate, while lime is deposited by this change. In practice it is usual to strew the gypsum on plants freshly covered with dew or moistened by rain. This method is justified by the fact that a solution of gypsum is rapidly formed on the wet plants, and dropping down from these into the soil, it becomes almost immediately active in proximity of the roots.

§ 11. How is the nutrition of pot-plants effected?

The points which are of greatest importance in dealing with the manuring of crops in the open are of secondary importance in speaking of pot-cultures, unless in some exceptional case there should be, by some gross neglect, an actual deficiency of some food substance. The various soils which are used for potting purposes all contain at least sufficient food matter for the growth of the plants; but even the richest can only last for a certain time in the very limited space of a pot. In all cases a renewal of the soil has to take place by re-potting, and it is the gardener's object to select the proper time for this proceeding.

If, therefore, the root of a pot-plant finds in every soil the necessary quantity of food substances, the usual choice of heather soil, leaf or peat mould, loam or turfy soil, for different plants, is not made because one kind of mould contains food substances which are absent from another, or because some substance is present in greater quantity in one soil than in another. Such differences exist of course in the different soils,[1] but are of little importance in the case of pot-plants, because usually the earth of a pot is renewed before the roots

---

[1] The following table of Loges gives a comparison of the amount of the various nutritive substances contained in a leaf-mould from various sources. The dry weight of these moulds contained—

	Willow (*Salix capraea*). Per Cent.	Aspen Poplar (*Pop. tremula*). Per Cent.	Scotch Fir (*Pinus silv.*). Per Cent.	Pine (*Picea excelsa*). Per Cent.
Nitrogen	1.46	0.97	0.95	0.81
Fats	4.08	9.49	8.72	12.01
Carbo-hydrates	57.92	45.14	44.36	46.23
Fibres	22.97	33.19	39.16	32.10
Potassium	1.60	0.31	0.33	0.06
Phosphoric acid	0.32	0.22	0.10	0.12

have exhausted the nutritive substances, and because we can renew these by liquid manures during the life of the plant, and often increase their amount above the original supply of food matter.

In the case of pot-plants, the choice of the various soils depends upon their physical properties, upon their power of retaining water, of absorbing nutritive substances, and allowing a sufficient amount of air to enter for the aëration of the root.

With regard to the latter, our cultivated plants have very different requirements. As far as the development of the plants is dependent upon the soil, we may say that: **It is not the total amount of nutritive substances contained in the soil which determines the vigour of growth of a pot-plant, but it is due to the degree of concentration of the nutritive solutions in the soil, and to the intensity of its aëration.**

It is the knowledge of this fact, gained by experience, which distinguishes successful cultures from the attempts of beginners, who are always searching for rich mould. The "richness," *i.e.*, the amount of nutritive substance, we can easily supply to every flower-pot, but we cannot do the same with the physical conditions which are necessary for the active growth, and especially for the respiration of the root.

We have at present no scientific knowledge of the actual amounts of air necessary for our various plants, and we must therefore at present make use of the practical experience of gardeners. Now this teaches us that **a root can never have too much air, but often has too little.** Without taking into account those plants which are provided with aërial roots, we need only remember the exposed roots of trees on rocky slopes, walls, and roads, and think for a moment of the habit of many plants in our greenhouses of sending up delicate rootlets out of the soil into the air at the edge of the pot. All plants are also able to grow in moss or sand if the disadvantages of these media are remedied by abundant watering with nutritive solutions. As soon as we are able to satisfy the wants of roots with regard to water and nutritive substances, we may choose beads of glass or crumbled quartz instead of ordinary soil. Indeed, by timely changes of the nutritive solution, and by constantly renewing the supply of oxygen, we can cultivate plants for years in water itself.

This fact, which has been proved by numerous experiments, shows us how little we are tied to any special soil in the case of potting plants, if we only pay sufficient attention to the aëration of the soil. It is advisable, unless we know by experience that a certain species can stand a heavy soil, to choose in the first instance **some light kind of mould.** To these belong, in the first instance, heather soil, then follow the soils rich in humus, such as leaf-mould, and lastly loam and clay.

Besides the costliness of employing heather soil in all cases, the latter has the disadvantage of not absorbing nutritive substances, and of drying up very rapidly. In extensive potting, therefore, leaf-mould is more largely used, as there is less danger of the above-mentioned disadvantages, though in its pure condition it too is liable to rapid drying up. The mould resulting from the decay of leaves and twigs is therefore rarely used in a pure condition, but is mixed according to necessity with some animal manure and with loam. In what way this changes the physical and chemical conditions of the soil may be gathered from our preceding chapters.

As soon as we have realised that a considerable and permanent permeability of the soil is one of the chief necessities in pot-culture, we have settled the vexed question as to whether riddled or unriddled soils are most advantageous. By riddling the soil only the finer particles are used; this decreases the size of the interstices of the soil, increases the capacity of retaining water, and tends to clog up the soil, and thus cuts off from the roots the proper supply of air. These dangers are not counterbalanced by any advantages whatever, for the argument that a closely packed soil offers more nutritive material to the roots than a soil with larger particles, and therefore larger interstices, has no weight when we remember that by a single application of liquid manure more food material can be added to the soil than could possibly be contained in the additional particles of the mould. In all pot-cultures, therefore, **unriddled mould should be used.** But we have mentioned before that another point is of the greatest importance for pot-culture, and this one is very often sinned against. We must take into proper consideration the **concentration of the nutritive solution** which is offered to the roots. The water contained in the interstices of the soil contains a greater or less amount of nutritive substances which it has dissolved on its way to the roots, and constitutes therefore the nutritive solution. According to the amounts present in the soil, the solution will contain more of one substance and less of another, and will therefore be concentrated in various degrees. Now, every root needs for the proper development of the upper portions of the plant a definite saturation of the nutritive liquid of the soil. The plant will not die if this advantageous degree of concentration is not reached, but it will not flourish so well as it might do with a greater concentration of the nutritive fluid, other things being equal.

If the fault lies in the fact that the solutions are too dilute, the plant knows how to remedy the defect. The roots absorb great quantities of the solution, and the plant gives off the superfluous water through its leaves by an increased transpiration, retaining the dissolved substances. The general appearance of the plant betrays to some extent the method of its nutrition, as large pale leaves are formed. No sickening of the plant, however, is noticeable. The case becomes more serious if the concentration of the water contained in the pot becomes too strong, owing to a too rapidly repeated application of manures. The growth of the upper portions of the plant is then visibly retarded, the internodes are shortened, and the leaves become puckered owing to the shortening of the midrib and lateral veins, or are bent in various directions, spotted, and fall off at an early period. The roots themselves are short, thick, and bent up, and the newly-formed root-hairs are irregular and shortened, soon become brown and discoloured, and die away. As I know well from the many plants which have been sent me by gardeners, these appearances are usually put down to other causes, as gardeners are hard to convince that plants may be easily overfed. **This overfeeding of plants is greatly on the increase at present, and we shall prevent many losses by being more moderate in the manuring of pot-plants.**

We must not forget that the kind of soil and manure which is sufficient, and perhaps the best, for a certain species of plant may be much too strong for another one, and cause it to sicken. Ericaceæ, Myrtaceæ, and many Leguminosæ require a comparatively weak solution of nutritive matter, while highly concentrated solutions are beneficial to Cruciferæ (especially our vegetables), Resedaceæ, Cucurbitaceæ, Chenopodiaceæ, &c.

If we seek for sound guidance from the aspect of the plant as to the amount of concentration which it requires, we may take it as a general rule that plants with leathery leaves, with hard and narrow leaves, and with hard wood, require more dilute solutions than those with large, soft, and expanded leaves. The period at which the manure is added is also of considerable importance. During the period of leaf-formation all plants can do with the greatest amount of nutritive matter.

If pot-plants have suffered from over-manuring, the ball of roots should first of all be examined, to see if the latter are decayed. If this is not the case, the surface of the soil should be loosened, the pots should be washed to keep them as porous as possible, and the watering should be reduced, so that a large amount of air can reach the roots. If the plants must have water, add it in such quantity that it escapes by the hole at the bottom of the pot. Pure water only should be used, and the pots should remain in their ordinary position. If, however, the roots be injured by decay, then the plants must be re-potted. After the removal of the decayed parts the plants should be potted in a loose soil in small pots, be given moderate bottom heat, a close atmosphere, increased shade, and diminished water supply. If bottom heat is not available, the plants should be entirely screened from the mid-day sun, and during the heat of the day the upper portions of the plant should be kept moist by sprinkling, but the watering of the roots should be reduced to a minimum. The plant must rest as much as possible until the development of new leaves indicates the renewal of its activities. Then the patient may be removed to lighter quarters, and the water supplied to the roots may be gradually increased. The efficacy of this treatment I have had ample opportunity of testing.

§ 12. How do aërial roots nourish a plant?

This question is of considerable practical importance, now that the cultivation of orchids and aroids has become such a hobby. We must try to ascertain whether the aërial roots are organs by means of which the gardener can improve the nutrition and the development of his plants. This question must be answered in the affirmative; for the aërial roots constitute an absorptive system, and can therefore derive some benefit from an enriched supply of food material, *i.e.*, from manuring. The absorptive portion of the root consists here of a fine but tenacious silvery layer, which surrounds the root like a sheath (*velamen*), and will be well known to any gardener who has seen the strong aërial roots of a Vanda or Aërides. The genus of Stanhopea lends itself best perhaps to the study of aërial roots.

If a transverse section be made through the aërial root of a Stanhopea, the central vascular cylinder will be seen surrounded as in all roots by the thickened layer, previously described as the endodermis. This layer of cells is surrounded on the outside by the soft cortex, and the latter is enclosed by a layer of thickened cells, which represents the outermost layer, *i.e.*, the epidermis of ordinary roots, the cells of which would form the real absorbing surface, growing out to form the root-hairs. In aërial roots, however, this is not the case; for this layer does not limit the root on the outside, but is covered in by a tissue (very extensive in Stanhopea, but very delicate in some orchids), and it constitutes, therefore, an **outer endodermis**. The real soft cortex, therefore, is bounded in aërial roots on either side by a layer of endodermis. Outside the outer endodermis we find the root-sheath or velamen.

The cells of this sheath are fitted together without any intercellular spaces, and form a radially or longitudinally elongated parenchyma, generally strengthened by spiral bands running in the cell-walls. A point of special importance is the fact that the cells have often perforations leading from one cell to another. Such apertures exist not only between the cells in the inner portions of the velamen, but are also found in the walls of the outermost layer. In cells which are thus perforated, there naturally exist no liquid cell contents of the usual nature, but they are filled with air. It is the presence of this air which gives to the roots their silvery appearance. In consequence of the perforations of this root-sheath (which must not be confounded with the root-cap, although both are formed from the protoderm), it is able to function like a sponge; it sucks up in a moment as much water as it can contain, and condenses water-vapour and other gaseous matter which is contained in the atmosphere like any porous body. This **power of condensing vapours** and gases suggests the chief function of the aërial root under ordinary circumstances; it will use this power to nourish the root-system and the whole plant, if it can pass the condensed matter through the thickened layer of cells (the outer endodermis) which separates the root-sheath from the cortical tissues. The enormous power an aërial root possesses of condensing water-vapour is shown by an experiment with an aërial root of Epidendrum, 12 inches long. In an atmosphere saturated with moisture it absorbed in twenty-four hours an amount of water equal to one-ninth of its total weight.

The passage of water from the root-sheath to the succulent cortex is made possible by the peculiar structure of the outer endodermis. Speaking generally, this layer forms a protective sheath, built up of long thick-walled cells, without pits or canals; nothing practically can pass through their thickened walls. But a more careful examination reveals the presence between the thick-walled elements of many small thin-walled cells filled with protoplasm, and frequently arranged in rows. These will attract the liquid which is condensed in the velamen, and will easily pass it on into the cortex. The outer endodermis of the aërial roots unites, therefore, just as we have seen in the case of the endodermis of the ordinary roots, two very useful functions. Its chief function is to protect the aërial roots from drying up during periods of drought; but at the same time provision is made to conduct the substances which are absorbed during periods favourable to the vegetative processes into the more central tissues, whence they will be carried by means of the vessels to the leaves.

In some plants the root-sheath or velamen is thrown off

when the aërial root attains a certain age. The same may be observed to take place when such roots enter the soil, which indicates that the velamen represents the special organ by means of which the root can derive its nutriment from the atmosphere.

This being the case, the practical treatment of plants provided with aërial roots is readily deduced. In all cases, whether the aërial roots supplement the ordinary roots, or whether they alone exist, they must be protected and increased, as being the means of collecting and absorbing food material. Occasional sprinkling of the aërial roots with dilute nutritive solutions may be recommended as the most suitable means of manuring.

§ 13. How do ordinary roots obtain their necessary supply of air?

We have already dwelt several times upon the fact that in all horticultural practices the great respiratory need of the plant must be satisfied. Not only must the surface of the plant be in constant contact with the most essential constituent of the air, namely, the oxygen, but all the internal tissues, too, must be constantly supplied with oxygen, so that the necessary oxidation which constitutes the respiration of plants may take place.

It is true that every green cell of a plant under the influence of light is forming new organic substances from the nutritive sap and the carbonic acid of the air (**assimilation**), and in so doing liberates oxygen; but this amount of oxygen is by no means sufficient for the respiration. This can be gathered from the fact that in living green cells (cells of algæ, for instance), under favourable conditions of light and nutrition, the functions cease, and the movement of the protoplasm is suspended, if they do not receive renewed supplies of oxygen. A living cell will recover from this state of suspended animation, if not very long after it enters upon it it is again provided with oxygen. Should the oxygen be withheld for some time, the cell will become **asphyxiated**. During such a period of lack of oxygen the living cell can take oxygen from other organic or inorganic substances which may be at hand; but this destruction of various substances may become very disastrous to the plant, as we shall have occasion to see in dealing with the "over-watering" of pot-plants.

If we are to take any account of the favourable conditions for aëration of plants, we must first of all study the contrivances which exist in plants for the interchange of gases.

In the first place, we must mention the breathing-pores (stomata) which occur in all green organs of any considerable size. These pores, which occur especially on the under-surface, sometimes on both sides of leaves, belong to the external layer of cells (**epidermis**), and consist of two semilunar or kidney-shaped cells, facing each other with their concave sides, and attached firmly together at their pointed ends. They contain chlorophyll grains, which are generally absent from other epidermal cells.

As the pointed ends of these sickle-shaped cells meet, they leave an elliptical space between them, and this represents the chimney or passage by which the air has access to the interior of the plant. Every such passage opens internally into a large cavity in the tissues of the leaf which lies immediately beneath the epidermis, and is called **the respiratory cavity**. From this cavity a large number of small irregular passages run in all directions, formed by small spaces (**intercellular spaces**) which exist between the several green cells of the leaf. The external air can pass, therefore, into the respiratory cavity, and from this latter through the intercellular spaces to every single cell of the leaf. Gases can therefore pass directly into every cell or out of every cell. In submerged portions of plants the cells derive the necessary oxygen from the oxygen contained in the water.

In very delicate green structures, and in those portions of plants which are devoid of chlorophyll, the stomata are absent, while in some leaves they may number 180,000 per sq. inch. In tissues which are devoid of stomata the interchange of gases is effected by the entire surface of the organ or by other structures.

In older stems and roots these structures are the so-called **lenticels**.

These are small structures penetrating into the cortex, and

consisting of cork cells arranged more or less in rows. Between these cork cells the air has free access to the cortical tissues. To the naked eye these lenticels appear as minute warts on the surface of the stem, and during periods of long-continued rain they assume a white mealy appearance (*e.g.*, on alders in the autumn, on potatoes in a wet summer). This appearance is due to the excessive production and swelling up of certain portions of the corky tissue, so that the outermost cells split apart and the loose cork cells are pushed out of the pore in the form of a loose powder.

The rapid movement of air in the interior of stems is ensured by the presence in some plants of large irregular but continuous passages between the cells (in submerged plants and plants inhabiting swamps), while in others the older vessels, which form the centre of the woody cylinder, are filled with air.

How eager all plants are to obtain sufficient air we can judge from the behaviour of some which send up delicate rootlets out of the soil into the moist atmosphere of a greenhouse or forcing frame. In nature, too, this occurs in many plants rooted in swamps. Avicennia, for instance, in its natural surroundings of a mangrove swamp, sends up innumerable lateral roots, which stand out above the surface of the water, and are provided with lenticels for the purpose of taking in oxygen.

Gardeners will all know well enough the upward-growing roots of palms and Pandanus. The tips of these roots, which project above the soil, but almost always remain short, have a mealy appearance. Sometimes several inches of the root are evenly covered with this powdery substance; in other cases the mealy covering is interrupted by several rings of the original epidermis. The root-cap is shrivelled up to a small brown cap.

Those regions which have a powdery appearance are no longer provided with an epidermis, but are covered in by a loosely arranged tissue, the intercellular spaces of which communicate freely with the intercellular spaces of the cortex. This layer of "**spongy parenchyma**" is formed in the youngest portion of the root-apex by an increase in number and a rounding off of the cortical cells, which cause the root-tip to become dilated into a knob. Below this spongy tissue the usual sclerenchymatous ring, which forms a protective sheath to the roots, is missing. If such an upward-growing root is cut off and is hermetically sealed into the shorter arm of a U-shaped glass tube, by pouring mercury or by blowing into the other arm, air can be forced through the root-tip. This proves that the aëration of the whole root takes place through the mealy-looking tip. This means of aëration by specially modified roots, which are termed "**pneumathodes**," occurs in many species of Phœnix, Livistona, Cocas, Chamædorea, Pandanus, &c. In the last-mentioned palms the respiratory organs are found on thick horizontal roots in the form of small wart-like lateral roots, scarcely ¼ inch in length; in Pandanus the "pneumathodes" are very small warts on the large prop-like roots.

The explanation of this phenomenon, to which little attention has been paid, of roots being sent up out of pots, is to be found in the sensitiveness of these organs to gases,[1] which has now been established. The gas which produces the abnormal growth in this case is the oxygen of the air, which is driven out of the interstices of the soil by watering. The more it is watered, the closer the soil, and the damper the pot is kept, the more the roots will grow towards the upper regions of the pot, where there is more oxygen, and even grow right out of the soil. If the soil is only kept moderately moist, so that the roots can obtain the necessary supply of air within the soil, plants which usually direct their roots upwards will produce their pneumathodes underground. It is their desire for oxygen which causes the roots to grow towards the inner surface of the pots, and this points out the necessity of occasionally **washing the outside of the pots** to keep their pores open.

[1] Aërotropism.

CHAPTER IV

THE TREATMENT OF ROOTS

§ 14. How should roots be treated in transplanting?

IN the treatment of the roots of our cultivated plants, we must first consider what part the root plays in the economy of the plant, and secondly, whether it is of economic value for us. In the case of annuals, it is the rapidly growing absorptive and fixing organ; in perennials, besides absorbing the water contained in the soil, it serves also partially as a storehouse for reserve material, which the plant wishes to keep for the next vegetative period. In fleshy or tuberous roots, the storage function lasts for the greater portion of the life of the plant.

Its chief function is as an organ of absorption for the nutritive solutions of the soil. It is self-evident that, other things being equal, the development of the upper portions of the plant, especially of the assimilating leaves, will depend upon the amount of nutritive substances which are absorbed. Conversely, the greater development of the leaves will result in a greater production of organic matter (assimilated substance), and therefore more of this plastic matter will reach the root system, and supply it with the means for producing new ramifications.

(a.) A Root System which has been considerably Pruned.

The above-mentioned reciprocity must always be taken into account. If sickly plants with few leaves, or no leaves at all, are able to form new roots, and if, on the other hand, plants with feeble or damaged roots are able to produce new leaves, it is obvious that such a growth must take place at the expense of reserve material which is stored up in the main axis of the plant. This will only take place, however, when an increase of temperature stimulates the plant to increased activity. This stimulus, which is so often applied in pot-cultures, may be directly injurious if the newly developed organs, which are always somewhat weakly at the commencement, are not sufficiently cared for. This is especially the case with recently formed roots of a sickly plant. As a rule, the pot is kept too moist for the slowly and sparingly developing root system, and as a considerable amount of organic matter is undergoing decomposition in the pot, a dearth of oxygen soon occurs in the soil saturated with water, and the young root-tips decay. In the case of plants which are producing new leaves from reserve substances, it often happens that those plants which are just recovering are placed side by side with and under the same conditions as healthy plants, which require a large amount of transpiration. The new foliage is only able to respond to this great activity if it is aided by an energetic absorption of the root system, as is, for instance, the case in the normal development of leaves in the spring. This, however, is not generally the case with the recovering plants; consequently the young leaves shrivel up, and the plant will be entirely destroyed.

In the treatment of plants which are restoring their root system by the production of accessory roots, the first rule is to so reduce the work of the leaves that it is in harmony with the activity of the root.

This rule is not confined to pot-plants, but applies equally to plants grown in the open. The former may be placed during the period of root development in closed damp houses or frames, so that the amount of their transpiration is reduced, and therefore corresponds with the reduced absorption of the root system.

In trees and bushes which are transplanted the root system is always injured; the most apparent injury is the absence of the root-tips and of the absorptive region immediately behind them. In the case of such a reduction of the absorptive root-tips, it is evident that the plant would possess too large an amount of foliage if all the branches which had been formed were left intact. How can the root system, which has been damaged and cut in taking it out of the soil, absorb sufficient water for the full development of all its leaves? However much we may water the root, it will be of little avail; it may even be injurious to the plant, as the saturation of the soil with water may cause decay to set in at the cut ends.

In consequence of what has been said in the previous chapters about the function of the delicate root-tips, we must emphatically contradict the view which is still held and acted upon by some, that in transplanting trees and bushes the branches should be left unpruned.

In support of this view it is often mentioned that pot-plants, in which the root system has been considerably damaged by transplanting, need not be pruned in. This argument, however, though it is actually true, is fallacious, for pot-plants can be, and are, placed after transplanting either in a shady place, or in the moist atmosphere of a frame, which reduces the amount of their transpiration in accordance with the reduced absorption.

Transplanted woody plants must therefore have their crowns reduced. It is only a question as to how the pruning should take place, so as to assist as much as possible the speedy formation of new roots. If we assume the roots to be properly pruned, the production of adventitious roots depends upon the excess of food substance formed in the leaves over their consumption. This excess will find its way down the stem into the root system. Again, other things being equal, the amount of assimilated food matter available for the roots will depend upon the rapidity of development of, and upon the amount of work done by, the foliage. The more rapidly, therefore, we can produce a large number of actively functioning leaves, the sooner will the stem be in a position to pass down material for the formation of new roots.

But with regard to a rapid and strong development of leaf surface, the several buds of a branch behave very differently, and we may take it as a rule that the uppermost buds of every branch are the first to develop and produce the largest amount of leaf surface. Now, for the purpose under consideration, we require a large amount of foliage at the earliest possible period, and we ought therefore to leave the ends of all the branches intact, *i.e.*, we ought not to prune at all. But opposed to this requirement is the necessity to reduce the whole system of branches to a certain extent, as otherwise, in the course of the summer, some branches would die, because the damaged root system is not able to supply all the water which would be needed by the entire system of branches. But often those branches which die off are just those which are most essential for a regular growth of the crown, and therefore are an irreparable loss to the tree. Both requirements must therefore be taken into consideration, and this is done by leaving intact some branches which are essential for the formation of a good crown, and reducing the intervening branches to one-half or one-third of their length, according to the amount of damage of the root system. In this way we secure the development of a number of leaves at the ordinary time, and consequently a certain amount of new assimilated food substance. For these few branches the amount of sap absorbed is quite sufficient, and the development of adventitious roots begins near the cut ends of the rootlets before the buds of the pruned branches, which only form a small amount of leaf surface, have begun to open. The absorptive organs will therefore have increased in proportion to the new leaf surface.

The present method is to cut back all the strong branches, and to confine the production of leaves to the lower, less vigorous, and later developing buds. Consequently, the unfolding of the leaves takes place at a later and hotter period, when the leaves will remain smaller, but in spite of that will require more water for their transpiration.

The first-mentioned method of "**partial pruning**" deserves the preference for transplanted trees and bushes.

It is of course taken for granted that the larger cut surfaces will be covered with tar or wax, and that the stems themselves will, if necessary, be protected from evaporation by a covering of moss or straw.

(*b*.) *A Root System from which only the Delicate Roots have been Removed.*

In discussing the treatment of roots in transplanting, we have purposely distinguished those cases in which the older roots have been removed from those in which the younger and more delicate roots only have been injured, for the methods of healing are different in the two cases. The younger a root, the more easily will it heal, and the more copiously and more rapidly will it produce adventitious roots. This can be very readily seen in trees which are taken up again a few years after transplantation. Around the wound produced by the cutting of a large root, there will be seen a ring of delicate adventitious roots, some of which will have become stronger roots. The adventitious roots, produced from younger structures, are perhaps not equally numerous, but their dimensions are so perfect that they are often not to be distinguished from the undamaged roots of the same age.

A root system, therefore, which has been pruned back considerably will take a longer time to produce a sufficient absorptive system, the stem therefore will have only a small supply of water, and will require a treatment tending to reduce the transpiration for a longer period than a plant in which the delicate roots only have been cut.

It is therefore most advantageous in pruning roots to avoid if possible the cutting of the older roots.

(*c*.) *The Treatment of Roots in Re-Potting.*

The necessity for re-potting occurs either when the root system is diseased or when the soil has become exhausted by very active growth.

In the case of diseased roots, the above-mentioned precautions with regard to the upper portions of the plant should be remembered. One is forced on these occasions to cut back the roots to the entirely healthy portions, and therefore to cut back to the old wood; and this must not be done sparingly. For if the smallest decayed portions are left, there is great danger, should the watering be injudicious or the choice of mould faulty (too heavy), of continuing and increasing the decay. From the infected centre liquid products of decomposition are sucked up into the healthy tissues, cause the decay of the stem, and in the end bring about the death of the plant.

It is essential to use very loose moulds and small pots for plants with sickly roots.

First of all, no attention need be paid to a possible dearth of nutritive substances which might occur after the plant has made new roots. Such a want can easily be met by a subsequent transplanting. At present, the chief aim is to produce as soon as possible a new supply of roots, and this necessitates the preparation of a moderately moist soil, which can be kept permanently well aërated. This even moisture must not be brought about by repeated watering, but by a preservation of the water contained in the soil in the first instance, and which must be prevented from evaporating. This is most frequently done by immersing the pots in a bed of cinders or sand, cocoanut fibre, sawdust, tan, or other porous substances, which do not easily decompose. It will be found of great advantage if such a bed can be heated. Amateurs growing plants in an ordinary room will find it advantageous to sink the pot in a larger one containing river-sand.

River-sand is in all cases preferable to sand derived from pits, as the particles of the latter are always more liable to adhere together, owing to the presence of clay, loam, and iron. If river-sand is not obtainable, the pit-sand should always be washed before use. High temperatures, which would be very beneficial in frames or green-houses, should be avoided in the dry atmosphere of a room. Sickly plants should receive more broken potsherds below the soil than healthy plants, as they require a better and more rapidly drained soil.

The re-potting of a plant with a healthy ball of roots is necessary when the soil becomes exhausted. A growing plant often requires a larger amount of soil, the old soil being entirely used up by the network of rootlets. The latter are so closely pressed against the inside of the pot, that some are always damaged in removing the pot. Now, it has become the general practice to greatly increase this injury by ripping up with a pointed stick this dense felted mass of rootlets and tearing it away. Experience has proved the utility of this practice if carried out moderately. If this felt-work of rootlets is left intact, and the whole ball of roots is immersed, just as it is, in a larger pot, the earth at the centre has first of all not been renewed; and secondly, a tremendously dense mass of roots partially damaged (by the removal of the old pot) is replaced in a new, moist soil, containing a large amount of humus. This dense mass of roots requires a very large amount of oxygen, which it could formerly obtain, being pressed against the side of the pot; but now the conditions have changed. In the new soil, which is richer in decomposing organic matter, and therefore requires for itself a larger amount of oxygen, a number of the pores are blocked by water, which absorbs the products of decomposition, especially the carbonic acid. It is this water which surrounds the felted mass of roots. Consequently the injured portions of the root system often begin to decay.

Secondly, the root-tips which remain pressed together retain, in the first instance, the direction of their growth, and only slowly bend towards the sides of the pot, whereas it is most advantageous for them to push as soon as possible strong root-tips towards the better aërated region on the inside of the pot. It is therefore better to remove a portion of the old network of roots, and so to open up a passage for the rapidly growing new roots to the side of the pot. Especially in the case of herbaceous plants will it be noticed that those with a loosened ball will most rapidly re-establish themselves in the new pot.

With regard to the mixing of a mould for pot-plants, we may adopt the general rule that it must be coarse, but not so light as to cause any danger of drying up. This is especially to be borne in mind in the case of plants grown in dwelling-houses.

(d.) *The Treatment of Roots in Transplanting in the Open.*

In the case of short-lived herbaceous plants grown in the open ground, the treatment of the roots is very simple. In pricking out the seedlings, the young tap-root is always damaged, and consequently the lateral roots soon make their appearance, so that the number of the absorptive organs is increased. We are more concerned with the woody plants which the gardener cultivates for commercial purposes or for his own use. In both cases it is necessary in their culture to pay attention to the portability of the plant. This is only attained when the whole root system occupies a very small space. Trees which are left to themselves develop where they are first grown a long tap-root, by which they are best fixed to the soil, and absorb from the deeper layers of the soil the necessary food material. The development of these deep-rooting plants is a relatively slow one, but correspondingly lasting. Trees which have been transplanted in their youth, and have therefore suffered an injury to their tap-root, produce near the wound several lateral roots, which grow on unhindered in a horizontal direction without much branching. To take up such trees requires considerable labour. Besides, with such diverging roots they are very troublesome to plant, as exceedingly large holes must be prepared for their reception. The general remedy for this is to saw off the long lateral root, so as to reduce the root system to a more convenient size. In doing so, however, the old wood is cut, and only few adventitious roots are formed, and their development is slow and weakly. The few greatly reduced roots give to the tree only a feeble hold on the soil, and produce only few absorptive organs, and cause therefore a considerable interruption of growth, which prevents the tree from attaining any considerable development, and causes premature death.

A tree intended for transplantation must have a root system consisting of a great number of short branches provided with many rootlets, so that the whole absorptive system is limited to a small area.

The large number of rootlets represents the greatest possible area of absorption, and at the same time a network which will take firm root of the soil, and cause the rapid formation of a ball of roots, which enables the transplanted tree to establish itself.

Such a much-branched root system, forming a nest or ball

of roots, can be systematically obtained by pruning the roots from the very commencement. This demonstrates the **necessity of repeated transplanting of young** trees which are intended for sale or for future transplanting. The treatment of young wild stocks begins, indeed, with the **pricking out** of the seedlings. A discussion of the technicalities of this process would be out of place in this book, but we may mention the most suitable time for this process. In countries with a mild winter, like England, November may be recommended, whereas in Germany March is preferable, as this prevents any damage to the young plants by the severe frosts, and the cut surfaces are soon healed by the recommencing activity of growth.

Some woody plants have the tendency to produce only few and sparingly-branched lateral roots, which resemble the taproot in growing almost vertically down to a great depth; in this case it is advisable to repeat the above-mentioned process. All the roots must be pruned, and the same must be done for the lateral branches of the stem. Conifers and other evergreen plants must also be transplanted at the end of their year's growth. But as this ceases at an earlier period than is the case in plants which lose their leaves, we may proceed to transplantation at the end of the summer in localities in which a late drought need not be feared. Thus the plant has the opportunity of forming new roots before the winter commences.

To preserve the moisture of the spring, and to keep the soil very porous, beds with such young plants should be covered with loose straw, moss, fibre, or short manure.

Seeing that the formation of long roots must be prevented, so as not to have to cut into the old wood, it is essential that young plants should be transplanted every year, and that (according to the species) the larger roots of the previous year be reduced to about two-thirds of their original length. At the commencement of the last third of the length there is sufficient tendency to form lateral roots so as to ensure a considerable number being formed, and the cut end is thin enough to be very soon healed over. Large cut surfaces are dangerous, because they are less rapidly covered over, and consequently are easily attacked by Fungi, which will cause decay.

It is from a dread of such consequences that the practice has grown up of not pruning the roots at all in transplanting, but of coiling them up spirally. This practice is further strengthened by the observation that roots which have been bent tend to produce numerous lateral roots immediately above the bend. This observation holds good for seedlings at least, as can easily be proved by growing them in water. When one of the long roots reaches the bottom of the vessel in which the plant is grown, it becomes bent, and lateral roots soon make their appearance.

Still the above-mentioned practice is to be condemned, as we will surely do after inspecting the roots a few years after transplantation. Generally the tap-root will be found again, only spirally coiled, so that its apex, which bears the rootlets, is closer to the surface. The older portions will not have produced any branches, and a number of those found near the apex will never be so great as in the case of a properly pruned root. We must also not forget that the nutrition of the plant cannot be so favourable under these circumstance, as the water has a longer course to take through the spirally coiled root. Now the bending of any organ always retards the passage of the sap; the shorter and more direct, therefore, the passage is, the more advantageous will it be to the plant.

Older plants, even when well cultivated and nourished, are always at a disadvantage as compared with younger ones, because both the sap and the plastic substances have to pass along much greater distances (sometimes no longer very passable) before they reach the final goal, the growing points of roots and branches. This causes the frequent dying off of the tips of horizontal branches of old trees.

As a matter of fact, therefore, the nutrition of plants with bent roots is less favourable than that of plants with pruned roots; besides this, the transplanting of the former is always more difficult.

We must not close the chapter on the treatment of roots without dwelling upon the important fact that **freshly transplanted trees and shrubs are more sensitive** than the untouched ones.

Generally speaking, the roots are more sensitive than the

stem and branches, owing to their more delicate tissues, and to the larger percentage of water of the former. The branches may, indeed, be made more resistant by certain injuries to the roots. Several cases are known in which the branches of fruit-trees which were transplanted in the autumn were less damaged by frost than those of trees which had remained in their original positions. Such a phenomenon is, in all probability, due to the fact that the branches contained less water, as the transplanting, by damaging the many root-tips, would interrupt the absorption of water, and consequently stop the growth of the branches, and accelerate the ripening of their wood. Under normal conditions the shoots of apple-trees in a heavy soil will go on growing until stopped by the frost.

The greater liability to damage by frost in the roots of trees transplanted in the autumn is not due to any greater sensitiveness of the root, but to the greater porosity of the soil, which enables the frost to strike to a greater depth. The earth around a newly-planted tree is much more porous than the natural soil, which contains more water and allows the air to circulate less freely; and this is just what protects the roots from frost. Hence, in transplanting in autumn, it will be found advisable to heap up the earth one or two feet above the soil around the tree, and open up the heap in the spring.

It is the porosity of the soil, which only settles down after many months, which causes trees transplanted in the spring to suffer from dryness. It is therefore always essential to wash the soil well into the root system. We should not be stopped from doing this by the fact that the soil is already very wet, or that heavy and continuous rain is falling at the period of transplanting. It is of prime importance that a layer of fine earth should cover in the rootlets as soon as possible, and this can only be done by washing it in.

The practice of placing the roots in water previous to planting cannot always be recommended. The tissues absorb an enormous quantity of water, and have afterwards to return to a soil with a very much smaller supply. Experience, too, teaches us that plants with great turgidity suffer most when a dearth of water arises.

In the case of freshly-transplanted trees, again, **manure of any great strength should be avoided.** The very highly concentrated solutions of nutritive substances which are formed in the soil under these conditions cannot be utilised by the injured root system. We always supply the roots as directly as possible with some compost, if we do not fill the entire hole with it, and this is quite sufficient to see the tree through the first year. If further material is available, a little decomposed manure may be spread over the surface of the soil. This will gradually become washed free of its nutritive substances, which will find their way into the soil below. We do not, however, wish to say that no manures should be put into the trenches. If the soil is very poor, the addition of manure (especially of animal dung) will probably be very beneficial; but it should be so placed that the roots will only make use of it in the second year.

In manuring trees which have not been transplanted, mineral manures are often preferable to animal manures, because the decomposition of the latter takes place very slowly in the soil, which is usually very dry under the trees. It is also more difficult to apply stable manures, and the effect produced by them is not always the desired one. Supposing the trees show a tendency to produce strong branches, and you wish to produce fruit-buds, animal manures are of much less use than the addition of phosphoric acid alone in the form of ground Thomas phosphate, &c. If the tree does not form proper woody shoots, a manure rich in nitrates and potassium should be applied.

For large trees it is best to introduce the manures into trenches dug round the tree.

Such a trench should be made about the breadth of a spade, and carried round the tree at about the same distance as the longest branches. It may even be closer up to the stem, but it should be made so deep as to reach the network of roots, the younger ones of which may be cut through. In the place of the excavated earth, compost, decayed manure, or, in absence of the latter, ordinary earth mixed with fresh dung, horn shavings, or other insoluble refuse, should be placed. The best effect is produced by rich compost.

The advantage of this method would be a decrease in the development of branches, owing to the injury to the roots, and the formation of numerous adventitious roots in the neighbourhood of the cut roots, and finally, an even and adequate nutrition.

There need be no fear of adding fresh animal dung in this case, as it always takes some time before the adventitious roots are formed, and have increased so as to be in contact with the dung. If liquid manure is to be used, it will suffice to dig a number of small holes in place of the trench. The roots will be produced most profusely in the neighbourhood of the centres of food substance; they have partly the tendency to grow towards the source of food matter (*trophotropism*), and partly to grow in the direction of greater moisture (*hydrotropism*), if the water is unequally distributed.

Turf under the trees should be avoided, as the roots of the grass absorb the rain-water, and in dry seasons may even draw away water from the deeper regions which should be reserved for the roots of the tree. In periods of great drought do not depend upon the protection of the turf, but remove it, water copiously, and replace the sods, roots uppermost. Even if no water is at hand for watering, lift up the turf, dig up the ground lightly, replace the turf with the green side downwards. The larger openings in the soil prevent the rising of the water by capillarity from the deeper regions of the soil, which retain what little water they have for the roots. The inverted sods, on the other hand, prevent or weaken the suction which the dry atmosphere exerts on the air contained in the loose soil.

CHAPTER V

THE STEM

§ 15. What is the structure of the stem?

THE root passes above the ground into the stem, which presents itself to us as an axis bearing the leaves, or in certain comparatively rare cases, undertaking the function of the leaves itself.

The external development of the stem axis varies very much in size, duration, and in the composition of its tissues. An axis which only bears flowers is spoken of as a *scape*, the leaf-bearing axis is called a *culm* or *haulm* (in grasses), a *stalk* (in herbaceous plants), or a *stem* (in trees). In some plants the main portion of the leaf-bearing axis remains underground, and only the leaves and the tip of the axis are pushed up above the soil. This, however, is not the place to discuss the general morphology and anatomy of the stem, as the physiologist and gardener are chiefly concerned with the function of the various organs, and their place and importance in the economy of the plant. From this point of view, it is sufficient for us to remember that we have already termed the leaves the chief seat of formation of organic substance, and to observe that the network of veins which is marked out on the leaf surface collects into considerable bundles which pass down the leaf-stalk and are continued down the stem.

If we follow the course of these bundles from a young leaf into the stem, we see that it is these leaf-trace bundles which form a ring of bundles in the stem, and which by their further development form the woody cylinder of the older stem. The number of bundles which pass from a leaf into the stem, and which we have already termed the fibro-vascular bundles, varies in different plants (3, 5, 7, &c.), as can readily be seen from the leaf-scars on the branches of our trees. The darker, knob-like processes which are seen on the smooth surface of the leaf-scar are the broken ends of so many vascular bundles, which passed from the stem to nourish the leaf. Each of these bundles can be traced down in the stem, passing several older leaves, until it becomes more and more delicate, and finally splits up between the bundles of the older leaves, its elements passing into those of the lower lying bundles.

If we examine a branch which has been cut across in the region where it is beginning to get woody, we will notice in the middle of the cut surface a soft parenchymatous tissue, which we call the **pith** or **medulla**. This is surrounded by a firm, white, and radially divided ring termed the **wood-cylinder**, the radially running lines being the **medullary rays**.[1] The wood-cylinder consists of thick-walled elongated cells, which interlock with their pointed ends, and are termed the **wood-fibres** (*libriform fibres*). Between these are found the actual **wood-vessels**. The medullary rays, however, consist of parenchymatous cells, which are often elongated in the radial direction, so that the passage of water within the medullary ray from the pith to the cortex is retarded by fewer walls than if it were to pass vertically through the ray. These medullary rays, therefore, represent the most convenient passages in a radial or horizontal direction. The wood cylinder is surrounded on the outside by a thin layer of very delicate and easily-damaged cells, the **cambium ring**, which passes over more gradually into the external layers. These layers terminate externally in those portions of the stem which are still soft in a layer of tabular cells, which fit very closely together, and are usually devoid of chlorophyll. This layer can often be separated in the form of a thin skin, and is termed the **epidermis**. In the older brown portions of the branch, new cells will be found to be formed below the epidermis, which consist of brick-shaped cells, and form the **cork layer** which constitutes the protective sheath around the soft cortical tissue. But the tissues outside the cambium are still further differentiated. This can best be seen by scraping with the finger-nail a green

---

[1] See Fig. 2, p. 11. In this longitudinal section through the wood and the inner cortex, the medullary rays are indicated by the letter *m*.

shoot of the vine. It will then be noticed that only a little of the soft green tissue will be removed, and then there will be laid bare a number of firm whitish bundles of fibres running in longitudinal direction. These are the **hard bast fibres** or **stereom**. They consist of long spindle-shaped and thick-walled cells, usually without contents, but very flexible. They are united into strands running on the cortex, and form the material which we use for tying up plants. These bundles of hard bast run parallel to the vessels of the wood, and under the microscope it will be seen that in a young branch one such zone of bast lying in the cortex corresponds to each portion of wood lying between two primary medullary rays. Such a group of hard bast cells has often a semilunar appearance, the convex side turned towards the outside of the stem. Protected on the outside by this group of hard bast cells, we find a number of strands of delicate cells, which either have the same appearance as the cambium cells (*cambiform*), or which represent tubes which have been formed by the fusion of a number of superposed cells. These tubes are characterised by the fact that the transverse walls which separated the original cells have not entirely disappeared (as in the wood vessels), but are perforated after the manner of a sieve. These tubes have therefore been termed **sieve-tubes**, and together with the **cambiform cells** they constitute the **soft bast**, as distinguished from the above-mentioned hard bast. It is along this soft bast tissue that the organic **plastic material** (assimilated material) which has been formed in the leaves passes to the regions where it is required.

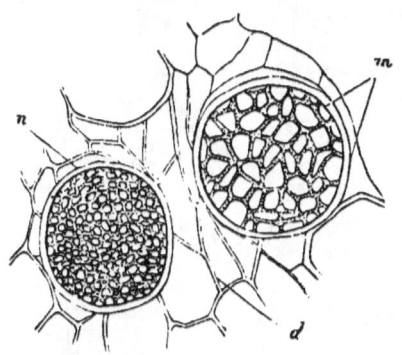

FIG. 10.—PART OF A TRANSVERSE SECTION THROUGH A VASCULAR BUNDLE OF *Lagenaria vulgaris*.
m widely reticulate, *n* closely reticulate sieve-plate. Each is the horizontal transverse wall of a sieve-tube (*after* DE BARY).

Fig. 10 represents a more delicate and a coarser sieve-plate, forming the transverse walls of two superposed sieve-tubes. In Fig. 11 is shown a longitudinal section through such a sieve-tube and its sieve-plate. The pores on the left-hand side are still filled with the protoplasmic contents, while on the right hand the action of the alcohol has caused them to shrink and be withdrawn from the pores. They have, however, been hardened by the alcohol in the form of finger-like processes, which at one time penetrated the plate.

It is not without ground that we have emphasised the fact that these groups of soft bast run on the inside of the cortex parallel to the strands of woody tissue. They are, in fact, a part of these fibro-vascular bundles which traverse the stem.

With this knowledge we are now able to picture to ourselves the structure of a stem or branch about one year old. If we consider such a shoot to be placed vertically upright before us, it will be found to be an elongate structure consisting of many stories. Every story is marked by the insertion of a leaf. The chief form of brick in this tower-like structure is the parenchymatous cell, which can best be compared with the cell of a honeycomb. Of such cells the leaves and the shoot are mainly built up. Every parenchymatous cell is a small chamber, which shows at a certain time of its life a great activity in forming new organised material. Such activity can only take place to any extent if new raw material is constantly brought to it and the assimilated substance is constantly removed from these manufacturing cells. For this purpose special passages are formed, the fibro-vascular bundles, which consist of two chief portions. One part consists of

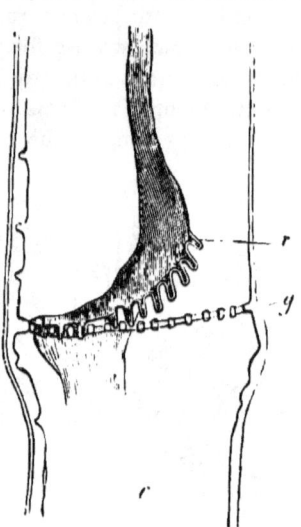

FIG. 11.— LONGITUDINAL SECTION THROUGH A LARGE SIEVE-TUBE OF *Lagenaria vulgaris*.

Its protoplasmic contents have been contracted by alcohol into the strand *r*. The protoplasmic processes which passed through the pores *g* have been hardened, and have shrunk away from the sieve-plate (*after* DE BARY).

woody cells and vessels, and constitutes the wood or xylem; the other contains the sieve-tubes, and is termed the **bast** or phloem. The first-named portion conducts the raw solution from the root to all the manufacturing cells, the second carries away the organic material which has been built up in these cells to all the places where new cells are to be formed.

In the leafy expansions we see the conducting bundles forming a delicate network, which penetrates the whole parenchymatous mass and collects in the midrib into the more compact bundles.

Each collection of bundles passes from the leaf into the stem, runs down it through four or five stories, and then unites with the bundles coming from a leaf at that lower level. In this manner each leaf-trace bundle runs for a time between those of the lower leaves, and at whatever point of the stem we make a transverse section, we shall always find the conducting systems of several superposed leaves.

The bundles which come from the leaves appear as it were let into the parenchymatous ground tissue, of which the first year's axis is mainly built up. As the bundles of dicotyledonous plants and of conifers are arranged in a ring (Fig. 12 A), they form later on a continuous ring which separates a central tissue from the external regions. This central tissue is termed the pith or medulla (M), the outer zone is called the cortex (R). If the fibro-vascular bundles after their entrance into the stem do not all run at a certain distance from the periphery, but first bend in towards the centre of the stem in a gentle curve, and then bend back towards the periphery, a transverse section would show no regular arrangement of the bundles. It would reveal the bundles as dark, firm, irregularly placed groups of tissue scattered through the ground tissue, and not separating a pith from a cortical region. This is the case in Monocotyledons, of which the stems of palms, Dracænas, and of the maize are easily procurable examples.

§ 16. What is cambium?

But it is not only in the course of the bundles that we notice a difference between Monocotyledons and Dicotyledons and Conifers. The composition of each bundle also differs. It is true in their first stages both are similarly made up of thin-walled cells containing much protoplasm, and are then called **procambial strands**; but during their later development they become differentiated in different ways. The differentiation of the procambium cells into wood vessels ($xylem$) (Fig. 12, A, x), and sieve-tubes or phloem (Fig. 12, A, p) begins at the periphery of each bundle, and proceeds towards its centre. But in the bundles of Conifers, and of most Dicotyledons, there remains in the centre an undifferentiated zone of the original thin-walled cells with their protoplasmic contents, and these retain and use their power of division, thus increasing in number. This persistent zone of cells, which are able to divide (*meristematic cells*), is absent from the bundles of Monocotyledons, and the several bundles cannot therefore increase in thickness during the succeeding years, as is the case with the vascular bundles of most Dicotyledons.

If, at the beginning of a new period of growth, long-lived plants require an increase of their conducting tissues, this can only be done in the case of Monocotyledons by the formation of new vascular bundles, while in the case of Dicotyledons the number of conducting elements can be increased within the already existing bundle.

For the purpose of answering the question placed at the head of this paragraph, we must remember that the stems of our woody plants have a ring of bundles inserted in their ground tissue But as all these bundles are arranged in the same way, so that the xylem or wood is turned towards the centre (Fig. 12, B, x), while the phloem or bast is turned towards the outside (Fig. 12, B, p, and b, b, b), the xylems form a cylinder which is only interrupted by the narrow medullary rays. The latter pass through the phloem too, and their cells have the faculty of fitting themselves to the tissue through which they have to pass. As long as the medullary ray runs through the wood, its cells are lignified, but in the bast the cells partake of the softer nature of this tissue. Hence the two portions of the ray may be termed the **interfascicular xylem** (Fig. 12, C, ifh) and **interfascicular phloem**, respectively (Fig. 12, C, ifp). By this adaptation of the

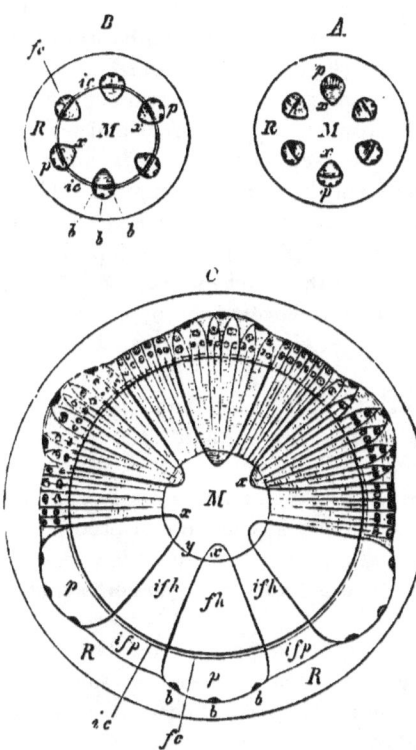

FIG. 12.—DIAGRAMMATIC REPRESENTATION OF THE GRADUAL FORMATION OF THE WOODY CYLINDER OF A DICOTYLEDON.

A, transverse section through the apex of a shoot; the several vascular bundles are arranged in a ring in the ground tissue, dividing it into central pith (M) and an external cortex (R); x, wood or xylem; p, bast or phloem.

B, transverse section of a somewhat older shoot. The lighter portion between the bundles represents the medullary rays. In each bundle there lies between the wood and the bast a layer of cambium, *fc* (*fascicular cambium*). This cambium is continued into the medullary ray, as *interfascicular cambium* (*ic*). Thus the cambium ring is formed.

C, transverse section through a mature shoot. The wood of the bundle (*fh*) and the interfascicular wood (*ifh*) form together the wood cylinder; *ic* and *fc* form the cambium ring. Outside this are the bast areas (*p* and *ifp*), the soft bast of the bundles being protected by the hard bast groups, *b* (*after* SACHS).

medullary to the surrounding elements, the various zones of the stem become homogeneous, and we can distinguish with the naked eye, an even woody cylinder and a correspondingly homogeneous ring or belt of bast.

But each bundle possesses, as we have seen above, a layer of cells lying between the bast (Fig. 12, *c*, *p*, and *b*, *b*, *b*) and the wood (Fig. 12, *C*, *x*, *ifh*), which layer is able to, and does actually, increase the number of its cells (Fig. 12, *C*, *fc*). This layer of cells representing some of the original procambium cells, which have remained active, and continue to multiply, we term cambium. The layers of cambium in each of the several bundles are also all equidistant from the centre of the axis. They do not, however, touch each other laterally, as a medullary ray runs between the bundles.

But these tissues of the medullary ray, where they separate two cambium layers, assume later on the same cambiform character, and thus form a bridge (Fig. 12, C, ic), which connects two adjoining pieces of cambium, and in this way a continuous cambium ring is ultimately formed (Fig. 12, C', fc and ic).

What appears in transverse section as a ring is, however, in reality a cylinder of cells which reaches through the whole length of the axis. The stem or branch, therefore, will now appear to be a column, the centre of which is occupied by a narrow but solid cylinder of pith. Around this we find a hollow cylinder of wood, which is covered in by a mantle of cambial tissue, bounded on the outside by the bast, still more externally by the cortex.

The view, therefore, which is still to be found in some of the older books that the cambium is a fluid, the so-called formative fluid, is erroneous. Cambium is not a formative fluid, but a formative tissue. Such formative tissues, which are composed of delicate closely-packed cells, full of protoplasm and plastic substances, and which have retained their powers of division, are termed *meristematic tissues*. **Cambium is therefore a meristem.**

So as to understand more clearly the position and the nature of the cambium, let us examine the transverse section of a one-year old shoot of the Laburnum (*Cytisus Laburnum*) which was cut in the month of May (Fig. 13).

The portion of the section figured contains only the essential tissues. It does not include the cortex, which would lie on the farther side of r, nor the pith, which would be found on the opposite side, *i.e.*, beyond m. We have therefore before us a portion of the wood cylinder which reaches to c: c represents the cambium ring, which is made up of thin-walled cells, fitting closely together, without intercellular spaces, and rich in protoplasm. These cells increase in number and become transformed towards the inside, *i.e.*, in the direction of nh, into new wood elements, on the outside into new phloem elements, while the median portion (where the letter c is placed) remains the actual seat of formation of new cells.

The tissue lying outside the layer of cambium consists chiefly of green cortical cells, in which are imbedded strands

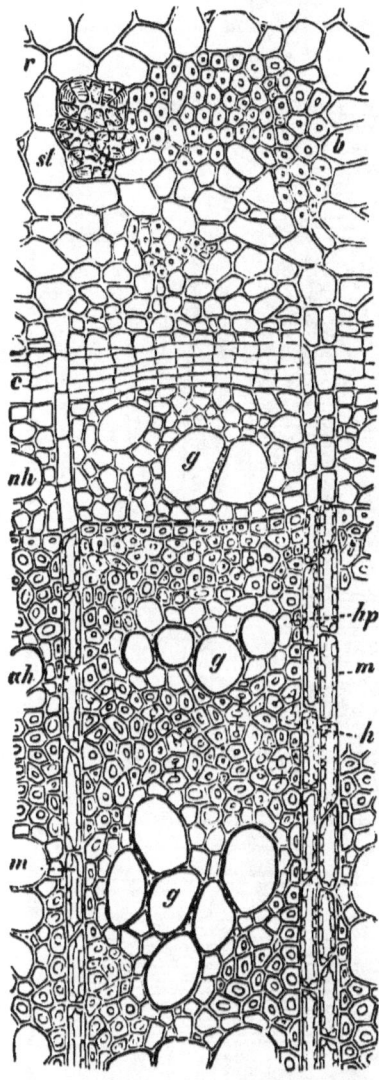

FIG. 13.—TRANSVERSE SECTION THROUGH THE WOOD AND BAST OF A ONE-YEAR OLD SHOOT OF *Cytisus Laburnum*, CUT AT THE END OF MAY IN THE SECOND YEAR (*after* LUERSSEN). *r*, parenchymatous cortex; *b*, bast fibres; *st*, schlerenchymatous cells; *bp*, bast parenchyma; *c*, cambium; *g*, wood vessels; *h*, wood cells (libriform); *t*, tracheids; *hp*, wood parenchyma; *g*, vessels; *m*, medullary ray.

of thick-walled hard-bast cells, *b*, and other groups of hard cells, *st* (sclerenchyma). The cells have the shape of parenchymatous cells, but have excessively thick walls, penetrated by canals or pits, and are very like bast fibres.

The woody cylinder can be divided into the firm wood of the previous year (*ah*), and the not quite so thick-walled wood of this year (*nh*) which has been formed up to the month of May. The elements which make up the old wood and the new wood are the same. We recognise the large open vessels (*g*) lying between the wood fibres (*h*), with their small lumen. These masses of wood are radially divided up by the lignified parenchyma, which makes up the medullary rays (*m*).

Fig. 14 represents a longitudinal section through the same piece of twig.

Here we recognise that the cambium layer (*c*) is built up of thin-walled cells with blunt ends, and which in their transition to bast cells (*bp*) still retain their original shape and size, but have become slightly broader and thicker-walled. On the other side, however, towards

THE STEM 103

y, they have already become elongated and pointed, and their pointed ends are penetrating between the upper and lower cells, so that the whole tissue becomes considerably firmer.

If we imagine a large number of such segments as Fig. 13 represents fitted together laterally, we have a picture of the transverse section of our branch with its cambial layer causing the increase in thickness.

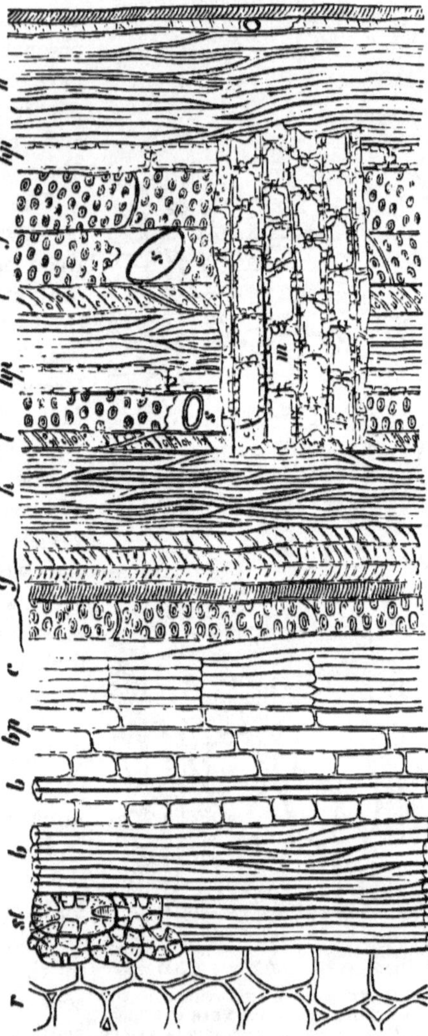

§ 17. **What is the function of the cambium in the ordinary course of growth?**

The ring of meristematic cells situated between the wood and the bast is placed in a very advantageous position for further growth. It can obtain from the adjoining wood cells as much of the unelaborated sap as it needs, while in the bast at a very little distance from the cambium are the sieve-tubes carrying down from the leaves the fully elaborated plastic materials. The

FIG. 14.—LONGITUDINAL SECTION THROUGH THE WOOD AND BAST OF A ONE-YEAR OLD TWIG OF *Cytisus Laburnum* (after LUERSSEN). *s*, the remnant of the transverse walls in the wood vessels. The lettering is the same as in Fig. 13.

cortex, too, and the medullary ray cells can take part in the nutrition of the cambium, as they are temporary storage tissues for reserve material (chiefly in the form of starch) with which they can supply the cambium in case of need.

This advantageous position of the cambium explains its continuous activity, which enables it to form on the inside new layers of wood cells (*splint-wood*), and on the outside new layers of bast, which push the older ones farther towards the periphery.

From this continuous activity of the cambium results the **growth in thickness** (the *secondary thickening*) of the stem.

The wood increases by additions to the outside, the bast by additions to its inner layers. Both regions of growth lie close together, and are only separated by the narrow band of cambium cells which are in process of division.

The new layers which are formed by the activity of the cambium, as they become older, grow and expand, and need therefore more room; consequently they press the cortex which lies externally to them towards the outside. But this is not done without resistance, for the cortex is surrounded by a belt of cork, the elasticity of which tends to retain the cortex in its original position. This tendency of the cork takes the form of a pressure, which makes itself felt as far as the cambium layer itself, and is one of the causes in determining the form of the new wood cells. That this is the case may be gathered from the fact, that immediately this pressure is relieved by cutting through the cork, the form of the wood cells is changed.

The increasing thickness of the stem finally ruptures the cork in various places; then new layers of rectangular cork cells will be formed within the green cortical tissues. The layers of cells outside these new cork layers are cut off from their sources of nutrition, they dry up and peel off in the form of scales. This is the mode of origin of **bark scales**.

Though the cork of a stem is very useful to the plant as a protection for the green cortex (for it only lets the smallest traces of water or gases pass through it), it is often very troublesome to gardeners. Trees grown on wet soil possess a very tough cork layer, which retards the formation of bark and the peeling off of the bark scales. This causes the pressure within the stem to be greater, the aëration of the cortex is reduced, and the thickening of the stem is kept back. In such cases the bark will suffer considerable decay, and a large development of lichens will take place on the trunk of the tree. These stems covered with moss and lichens are generally due to an excessive external pressure of the cortex, and should warn us to take some measure against it. These consist in scraping the stem or making longitudinal incisions as well.

The practice, which is gaining ground, of **scraping the stems** of our cultivated trees, is very much to be recommended as a means of preventing the excessive pressure of the cortex, and has the secondary advantage of preventing the old bark from harbouring insects and their eggs.

§ 18. **How does the stem as a whole perform its functions?**

We have sketched out in the preceding paragraphs the structure, arrangement, and functioning of the wood and bast systems, and of the cambium; it is now necessary to add a few words about the central cylinder of the pith. This latter consists of parenchymatous cells, which in a few families of plants is traversed by strands of soft bast, or even by entire vascular bundles—that is, by tissues meant for the conduction of formed substances. Such **medullary phloem bundles** are present, for instance, in the tuberous Begonias, which have to fill their storage tubers with reserve material for the leafless period of rest; they are absent in the ordinary Begonias. Medullary bundles exist, therefore, in plants which withdraw their reserve substances into the roots. From the fact that in many plants the pith may die off, split, dry up, and disappear, without affecting the stem, we may gather that the chief function of the pith must occur during the early development of the stem. As a matter of fact, experiments go to prove that the function of the pith is to swell up the axis, and therefore to accelerate the growth in length of the stem. If we cut a piece of a young stem or branch of the Elder (*Sambucus*), and place it in water, we see after a short time that the pith bulges out from the cut end, which indicates that by taking up water, it increases much more in length than the other tissues, especially more than the woody cylinder. If we remove the latter, the pith will increase still more in length, which shows us that it is its attachment to the wood which prevents the pith from increasing to its full extent. The rapidly elongating pith, however, drags the young ring of wood with it, as long as the latter is still soft and thin-walled. Besides this chief function, it has a secondary one, which often lasts until the stem is very old, and that is of storing away the reserve substances of the tree. In the winter, either a few rows of cells, or the whole of the pith, may be found to be filled with starch.

The stem, therefore, which is enormously developed in long-lived plants, appears in spite of its great dimensions to have only a subsidiary function in the economy of the plant. It is primarily a framework which bears the organs of chief activity, the leaves, the vascular bundles of which it connects with the single central vascular cylinder of that organ, which is second in importance to the plant—namely, the root. Like all organic machines, as distinguished from mechanical ones, the stem has the power of developing and completing its own structure to suit new demands. By means of its pith it can increase in length, and thus make room for new generations of leaves; its ring of cambium enables it to form new conducting tissues for the increased number of leaves. The conducting tissues are of two kinds. The wood contains vessels which lift the nutritive solution taken up by the root to the centres of assimilating activity, and carry them by delicate ramifications (veins) to the most extreme points of the leaves, the green cells of which have the power, under the influence of light, of building up from the carbonic acid of the air and from the salts of the soil new organic material which can be used for further growth.

Whatever material the leaf cannot itself employ at once for its own growth passes into the second part of the vascular bundle, the soft bast, and through the sieve-tubes of the latter into the axis. There the formed food substances pass slowly downwards through the bast, giving off laterally (both to the pith and to the cortex) whatever may be needed by these tissues. This stream of assimilated food matter passing down from cell to cell, and from sieve-tube to sieve-tube, regulates the growth of the cambium—that is, the yearly increase in thickness. In poor years, when the leaves are not very active, or when they are not well developed (*e.g.*, after attacks of insects), it may happen that the plastic material is used up on its way down, and nothing will remain for the cambium of the roots, which requires it as much as that of the stem. In such cases, the annual ring of growth tapers downwards, that is, the annual ring will only be found so far down as the formed material has been able to penetrate. In normal years, however, there is a surplus of food material. The tissues destined for its passage are not sufficient for the downward current of assimilated substances; it overflows into other tissues and there deposits the superfluous matter. In this case, the parenchymatous cells of the cortex, of the medullary rays, and of the pith become filled with starch, which remains in reserve for future use.

If trees are left to themselves, their upright trunk becomes more imposing by the death and decay of the lower branches. Leaving out of account the damage of the branches by herbivorous animals, the tree is able to **rid itself of its superfluous branches**. Either through the overshadowing of its neighbours, or through that of its own upper branches, the more delicate lower branches die off and are torn away by the wind.

The formation of a trunk is therefore advantageous to a tree, and called forth by the natural condition of growth and of its surroundings. In cultivated plants, however, it is not always advantageous nor necessary. This has brought about the cultivation of **dwarf-trees**.

CHAPTER VI

THE LEAF

§ 19. Which cells of the leaf are the most essential?

A MICROSCOPIC examination of the leaf shows us that it is mainly composed of green parenchymatous cells, which are so arranged as to form a flat expanded surface. Parenchyma is, as we have already stated, a tissue which consists of rounded or polygonal cells, the diameter in various directions being approximately the same, though the cells may sometimes be elongated to form a cylindrical structure. The walls are of moderate thickness, and the cells are, for a time at least, or permanently, filled with protoplasm, in which the vital processes of formation and transformation of organic matter take place. **The cells of the parenchyma are therefore the most important organs of the plant;** all the other forms of tissue are only of secondary importance. This is true of the thin-walled loosely packed cells forming the merenchyma (*e.g.*, the pulpy tissue of fruits), which is poor in protoplasm, and also of the sclerenchyma (hard tissue of nut-shells), the cells of which resemble parenchymatous cells, but are very thick-walled and entirely devoid of contents. The prosenchyma, too, consisting of long, pointed thick-walled cells, such as form the wood and the hard bast, is, functionally speaking, only of secondary importance. All these elements have secondary functions, and are not found in the simpler forms of green plants, such as Algæ and Mosses. Those which are thick-walled serve chiefly as protection and support for the parenchymatous cells.

The chief function of the green parenchymatous cell is the production of new organic material (*assimilation*), of which we have already spoken in passing. This formative process only takes place under the influence of light. The carbonic acid which enters the plant is broken up within the cell, and a portion of the oxygen is given off. This taking away and liberation of oxygen from the carbonic acid is termed a **process of reduction.** At the same time a very complicated process of transformation of organic substances is going on (*metabolism*), during which oxygen is taken up and carbonic acid is given off, which shows that a **process of oxidation** is also going on. The giving off of carbonic acid, due to the oxidisation of the organic substances (combustion of substance), is of the same nature as the breathing in of oxygen by animals, and the exhalation of carbonic acid, and is therefore termed respiration also in the case of vegetable tissues. This process takes place in darkness as well as in the light.

All parenchyma is not meant for assimilatory processes. We have already studied the parenchymatous cells of the pith, which rarely contain chlorophyll, and have a special function to perform. In other cases, the parenchymatous cells may have the function of a supporting or strengthening tissue. In such a case, the cell-walls become thickened in the corners where three or more cells meet, and the chlorophyll granules are very slightly developed. Such a tissue receives the special name of **collenchyma.**

The amount of assimilation which takes place depends upon the number of **chlorophyll granules** present in the cell. For it is these bodies which make use of energy latent in the rays of light, and perform the actual assimilation. This chlorophyll occurs in the phanerogamic plants in the form of granules or corpuscles, while in some of the algæ it assumes the form of plates or bands. The form of granule seems the most practical, for it offers the greatest amount of surface for the smallest amount of substance. In this way, within the smallest space the largest amount of light is absorbed, and "*light is life.*" The endeavour to absorb the greatest possible amount of light, which is apparent in most organisms, is made evident in plants by the phenomenon of **positive heliotropism**, *i.e.*, the faculty of placing their organs in such a position with reference to the source of light, that the incident rays will be most advantageously absorbed. Each chlorophyll granule consists of a protoplasmic framework, of a spongy nature, which is itself colourless, but is impregnated with the green colouring matter termed chlorophyll, which is soluble in alcohol. Sometimes there are only a few chlorophyll grains in each cell; generally, however, they are pretty numerous (sometimes more than 30), and they lie imbedded in the semi-fluid layer of protoplasm which lines the inside of each cell-wall, and is termed **primordial utricle.** In some cells of the leaf, which have a large lumen, a movement of this protoplasmic lining will be noticeable, carrying with it the chlorophyll granules. By this means they are able to wander in each cell towards the point which is best supplied with air and light. During the day they place themselves near the upper surface of the cells, and in the angles which adjoin the air-containing intercellular spaces (position of *epistrophe*), while at night they move towards the inner walls which adjoin other cells (*apostrophe*).

§ 20. **How are assimilating cells of the leaf protected?**

As we see that without chlorophyll granules a cell is not able to form any new organic matter, we must come to the conclusion that the most important and essential formative body in the vegetable kingdom is the chlorophyll grain.

We are therefore completely justified in saying that "**without chlorophyll there can be no organic life.**" For, as a matter of fact, the existence of animal organisms is dependent upon the presence of vegetable organisms. Carnivorous animals feed exclusively upon herbivorous animals, and the food of the latter, the products of the vegetable kingdom, depend upon the activity of the chlorophyll grain. The objection which might be raised, that the Fungi represent a large group of plants which may produce a considerable amount of matter without possessing any chlorophyll, does not hold good, for Fungi can only live on organic substances. If the latter still form part of a living organism, we are dealing with a **parasitic Fungi**; but if the substance is no longer connected with a living organism, and is in process of decomposition, we term the Fungi **saprophytic.** In either case, the substance is of an organic nature, and consequently the result (direct or indirect) of the activity of the chlorophyll.

After this exposition we shall not be surprised to find that nature has arranged these chief formative organs in such a

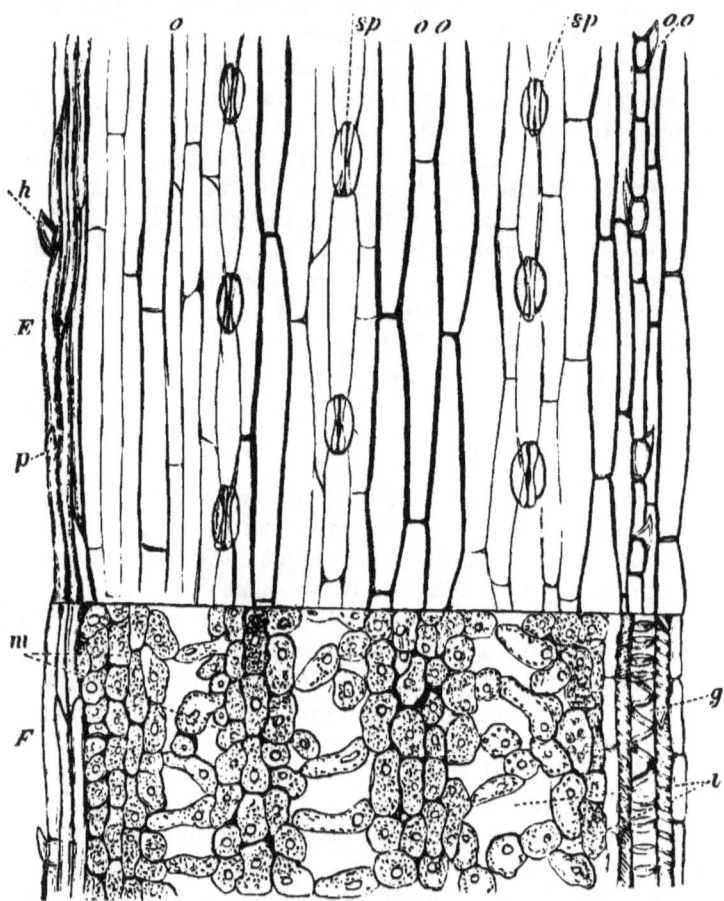

FIG. 15.—SURFACE VIEW AND SECTION OF THE LEAF OF THE BARLEY.
E, a piece covered by the epidermis; *F*, portion from which the epidermis has been stripped; *o*, long, thin-walled epidermis cells covering the mesophyll, which lies between two veins of the leaf; *oo*, thick-walled epidermis cells covering in the vascular bundles (*g*); *sp*, stomata leading to the spongy parenchyma, with large intercellular spaces (*i*); *m*, mesophyll cells; *h*, short hair on margin of leaf; *p*, prosenchymatous strand of cells strengthening the leaf margin.

way as to derive the greatest possible benefit from the smallest amount of substance. This is done by building the assimilating organs in such a way as to display the greatest superficial area.

This principle is carried out, in the first place, in the shape of the chlorophyll-containing bodies themselves, which are divided up into numerous small granules. Besides this, the assimilating cells are united so as to form thin expanded surfaces, the leaf-blades.

In these it will be noticed that the assimilating cells (Fig. 15), the actual green tissue (mesophyll) (Fig. 15, *m*), is protected on either surface by a layer of cells, which can often be stripped off as a colourless skin. This protective covering (epidermis) consists of closely fitting tabular cells, with a thicker outer wall, and generally devoid of contents (*o* and *oo*). These cells are covered in on their outer surface by a continuous layer of a substance which exhibits at times the character of cork, at times that of wax, and which renders the cells more or less impervious to gases and liquids. This layer is termed the **cuticle**; the more fatty matter this cuticular layer contains, the less liable are the leaves to wetting. The **bluish or whitish bloom** of many stalks, leaves, and fruits is due to a vegetable wax contained in or covering the cuticle.

The cuticularised epidermis does not, however, entirely close up the leaf. This would render the activity of the green mesophyll entirely impossible. For the chlorophyll-containing cells have the function of decomposing the carbonic acid when they are in the light; there must therefore be means by which the carbonic acid gas of the air reaches the interior of the leaf. This means exists in the form of a system of pores, provided with a mechanism for opening and closing.

Between the cells of the leaf-tissue or mesophyll are irregular spaces connected one with the other, which are termed **intercellular spaces** (*i*), and these are filled with air. Every green cell of the leaf can therefore take in gases from the intercellular spaces, and give off gases into them. The air contained within the leaf undergoes changes by the activity of the cells, and collects in large and definite spaces or **respiratory chambers**, which lie immediately beneath the epidermis, and above which the latter is interrupted. The openings of the epidermis are termed **breathing pores** (*stomata*,

Sp). They are formed by the splitting apart of two semilunar epidermal cells, which are formed specially for this purpose (Fig. 16, *Sp*, *sch*). These semilunar, or rather kidney-shaped cells, have their concave sides applied so that the ends of the cells alone are in contact, and a space, the actual pore, lies between them (*Sp*, *c*). If these kidney-shaped cells are flaccid, and are drawn out lengthways by the surrounding epidermis cells, the bend of each cell is lessened, and the intervening slit or pore is closed. Seen from above, the stomata will then have the appearance of a coffee-bean (*Sp*). If, however, the kidney-shaped guard cells become turgid through the absorption

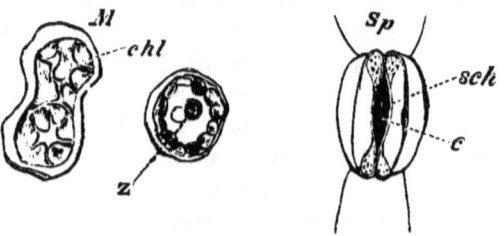

FIG. 16.—MESOPHYLL CELLS AND STOMATA OF THE BARLEY-LEAF.

Sp, a stoma; *sch*, guard cells; *c*, aperture of stomata leading into the respiratory cavity of the leaf; *M*, parenchymatous cells of a leaf, taken from near a vascular bundle; *Z*, nucleus; *chl*, chlorophyll grains.

of water, the line (*c*) in the above figure widens out to form an oblong canal. Through this canal a relatively large amount of air can enter the respiratory cavity or escape from it. If we consider for a moment that the leaf of an apple-tree has over 100,000 such pores to every square inch of leaf surface, we can form an idea of the ease with which an interchange of gases can be effected between the atmosphere and the inside of the leaf.

The structure of a leaf will become still more clear to us if we examine the transverse section of a leaf (Fig. 17). We will take as an example the leaf of the india-rubber plant (*Ficus elastica*), which is suffering from a disease very prevalent in plants grown in dwelling-rooms, and termed *intumescentia*. This disease consists in the appearance of numerous small

114   THE PHYSIOLOGY OF PLANTS

glandular protuberances on the under surface of the leaf. These are formed by the tubular elongation of cells (*int*), which in their normal condition have the appearance shown in *m*, and are therefore provided with considerable intercellular spaces. The diseased tissues on the under side of the leaf become more like the normal pallisade parenchyma (*p*) of the

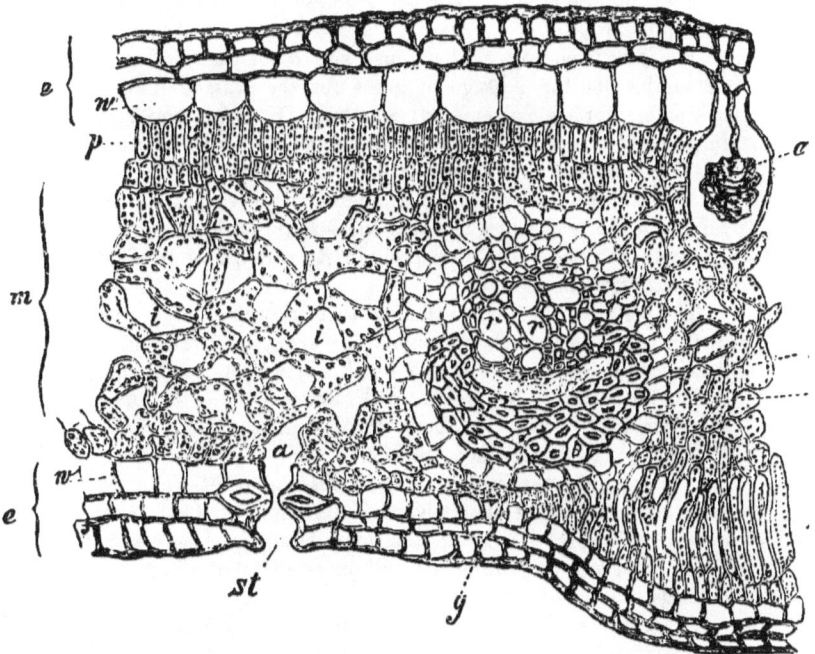

FIG. 17.—TRANSVERSE SECTION THROUGH THE LEAF OF *Ficus elastica* SUFFERING FROM OVER-WATERING.

upper surface of the leaf, which is covered in by a triple epidermis (*E*). Of these three layers, the outer one consists of very small cells, which are, however, covered in by a very thick waxy cuticle. The innermost layer of the epidermis is formed by thin-walled comparatively large cells (*w*), which must be looked upon as a water-storing protective layer. Some few cells of this layer contain curious structures of cell-wall substance (*cellulose*), which are incrusted with carbonate of lime. The part played by these grape-like crystals, which are termed cystoliths (*c*), in the economy of the leaf is at present still unknown. Such protrusions of the cell-walls with crystallised deposits have only been found in a few natural orders (*Urticaceæ*, *Ficoideæ*, *Acanthaceæ*, *Cucurbitaceæ*).

The firm covering of the upper surface of the leaf is not favourable for the passage of gases, for the passage of which the under surface is peculiarly adapted. Here we notice the large intercellular spaces in the tissue, which on that account has been termed the spongy parenchyma. The air contained in the intercellular spaces (*i*) communicates with the air contained in the respiratory chamber (*a*), and through the stomata (*st*) with the outer air, which can in its turn freely pass into the tissues of the leaf.

The water is supplied to the leaf through the veins, a small one of which is seen cut through transversely (*g*). The letters *r* represent the vessels of the wood, the actual canals along which the water flows. The path taken by the organised food material, which has been formed in the leaf, and of which there is a superfluous supply, is along the bundle-sheath (*sch*) down into the stem. *k* indicates the point where the cells begin to round themselves off abnormally, owing to the excessive supply of water; the intercellular spaces become obliterated, and they press the still more irritated tissue (*int*) towards the outside.

A structure similar to that of the leaf of Ficus would be presented by the leaves of most of our cultivated plants. Many of them, especially the herbaceous ones, have stomata on the upper side of the leaf, as well as on the under surface.

The functioning of the stomata will call forth our astonishment and admiration still more, if we reflect upon the fact that the guard cells, which are considerably thickened on their inner side, become more bent, and cause therefore an opening of the stomata, when they become turgid. But they only become turgid when the whole leaf is turgid, that is, when the roots supply the leaf with a large amount of water, and when there might arise some danger of a superfluity of water in the leaf tissues. In such cases, the plant protects itself against danger by enlarging the passages through which the water vapour can escape. On the other hand, the tissues in the inside of the leaf are protected from too great a loss of water through evaporation; for as soon as the guard cells lose water and become less turgid, they close the aperture which existed between them. Thus the plant regulates the amount of its **transpiration** according to its needs and requirements.

In many plants such safety-valves are still more largely developed. The tips and teeth of some leaves have very large stomata, arranged singly or in groups, and their guard cells are either rigid or sometimes quite destroyed. They lie immediately above a specially formed very thin-walled parenchymatous tissue (*epithem*), which reaches up to the epidermis. The vascular bundles of the leaf terminate in this epithem by a number of lignified cells or tracheids, often arranged after the manner of a brush. Now if the tissues of a plant contain an excessive amount of water, it is given off in these places in the form of liquid drops of water. Here, therefore, the stomata are **water stomata** or **water glands**.

This explains the phenomenon, which is well known to all gardeners, that in cool nights the tips and teeth of the leaves of some plants (*Primula sinensis, Calla œthiopica,* &c.) are quite regularly studded with small **drops of water**, and new drops will make their appearance as soon as the first are removed.

Just like the leaves, the green herbaceous stems are also covered by an epidermis perforated by stomata, so that the cortical tissues can be aërated. When the stem grows older, and the simple epidermis is replaced by a more resistant layer of cork, we find that below the stomata those more complicated respiratory organs will be formed which we have already discussed—namely, the **lenticels**.

The description of the epidermis would be incomplete if we omitted to mention its excrescences or **trichomes**. We understand by these first of all the hairs, which in their simplest forms are conical or tubular protrusions of the epidermal cells, as we have seen in the case of the root-hairs. Just as in the first stages of development, some cells of the epidermis remain small and meristematic, and give rise by dividing to the guard cells of the stomata, while others remain tubular, and constitute the ordinary epidermal cells, so some other cells which will not become guard cells also remain meristematic, and retain the power of growing out, dividing and branching, and represent the organs known as hairs. Sometimes a certain connection may be observed between the hairs and the stomata; in some petals, for instance, the formation of hairs increases as the formation of stomata ceases.

In the case of the formation of hairs, the plant again proceeds on the lines of increasing its superficial area. As long as the wall of these outgrowths is thin and not cuticularised, they will be able to absorb water and gases, and also to give them off; they will therefore be able to serve the purpose of transpiration. Later on, when the trichomes have grown to their full size, and have begun to form firm, cuticularised hairs, stellate, glandular, or woolly hairs, their function is a different one. They now represent a covering to the organ, and in the interstices of this covering they hold fast, very fast indeed, a certain quantity of air. It is only with great difficulty that the air can be driven out of such a felting of hairs, and it therefore is a very considerable protection against variations of temperature and of moisture.

Cold will take a long time to affect a leaf covered with such a felted mass of hairs, and rain will not be able to drive out the air from this covering, and will therefore not wet the leaf. But the greatest protection which such a layer of hairs affords is that it will greatly reduce the amount of transpiration from the leaf surface. The wind which passes over a bare thin-walled leaf carries away a great deal of water vapour, while it will scarcely set the air contained between the dense hairs in motion. The amount of transpiration, however, does not depend entirely upon the dryness and the movement of the air, but also largely upon the amount of light. The more intense the illumination, the greater is the evaporation. But in this case, too, the covering of hairs acts as a protective coating, for it prevents a large amount of light from reaching the leaf.

This explains why so many desert plants, and plants growing in sandy or rocky places and exposed to a powerful sun, are densely covered with hairs.

For horticultural purposes, too, we may deduce the rule that **plants with leaves densely covered with hairs require generally a large amount of light and little water.** We shall refer to this again in dealing with the watering of plants.

§ 21. How are the assimilating cells arranged within the leaf?

We have mentioned in passing that the cell of chief importance is the chlorophyll-containing cell; its function is to assimilate. These cells are so arranged in the assimilating organs that the chlorophyll grains can perform the greatest amount of work. The principle of increasing the superficial area is carried out to the fullest extent. The assimilating surface is protected by a well-developed and ingeniously devised epidermal covering.

We are now able to examine a little more closely the position of the assimilating cells themselves, and we shall find that here too we have a very advantageous combination of various contrivances, forming a most complete arrangement for ensuring the maximum of efficiency.

Most of the leaves are horizontally expanded, so that we can distinguish two surfaces, generally of different appearance, one directed upwards towards the sky, and the other downwards towards the earth. This latter surface is generally of a paler colour, which is due to the fact that the green cells of the mesophyll are more loosely arranged and are less rich in chlorophyll, and that the large intercellular spaces are filled with air.

The assimilating cells, however, which lie beneath the epidermis of the upper side of the leaf are very closely arranged, like a number of stakes driven in side by side, with their narrow ends towards the upper surface, and consequently with their long axis at right angles to the surface of the leaf. These cells form what is called the **pallisade parenchyma**. The loose, often horizontally elongated, tissue of the under side of the leaf is termed the **spongy parenchyma**. The pallisade cells, on account of the greater number of chlorophyll grains which they contain, and because the incident rays of light pass through the entire length of the cell, and illuminate each chlorophyll grain, are the most efficient assimilating cells. At the same time, as they have only their narrow head end directed towards the light, they are able to protect themselves from an excess of light.

The product of their activity is recognisable very soon after the light has been acting upon the tissues; it makes its appearance in most plants within the chlorophyll bodies in the form of **starch grains**. In some natural orders (*e.g.*, in the *Compositæ*), in the place of starch a soluble substance, inulin, is formed, which also consists of hydrogen, oxygen, and carbon, in about the same proportions as starch. If the water be withdrawn from cells containing this substance—as will be the case if the tissues are treated with alcohol—the inulin attaches itself to the cell-walls in the form of solid crystalline masses, the so-called **sphærocrystals**. Certain reactions, too, go to prove that new albuminous (nitrogen-containing) substances are formed within the chlorophyll grain; so that the two chief groups of the organic matter are represented within the chlorophyll granules.

If the pallisade cell cannot itself make use of the newly-formed organic material, it passes out of the cell during the night in the form of sugar and other soluble substances. These are either absorbed by the adjoining cells of the spongy parenchyma and conducted towards the base of the leaf, or they pass directly into the parenchymatous sheath which surrounds the vascular bundle. These bundle sheaths (Fig. 17, *Sch*) are distinguished, according to their contents, as sugar sheaths or starch sheaths, and form the actual canals along which the assimilated substances pass and rapidly reach the stem.

§ 22. How is the leaf developed?

If we examine the extreme tip of the stem, we find that it is composed of undifferentiated cells, fitted very closely together, with their delicate cell-walls, and containing a large amount of protoplasm.

These isodiametrical cells form the **primary meristem**. But very close behind the apex the tissues of the stalk begin to become differentiated, and first of all the cells of the outermost layer elongate and begin to form a tabular row of cells, which becomes the epidermis of the stem. Near the centre of the axis a number of rows of cells become differentiated by rapidly dividing by longitudinal walls, and forming therefore strands of tissue consisting of very narrow cylindrical elements. These strands, which go on dividing, are the rudiments of the vascular bundles, which occur scattered in Monocotyledons, but are arranged in a ring in Dicotyledons, where they will afterwards form the wood cylinder. The remainder of the primary meristem, which does not enter into the formation of epidermis or vascular bundle, remains as a parenchymatous ground-tissue, the central portion being called the pith, the outer portion the cortex.

In this first period of differentiation of tissues, it will be noticed that at certain points groups of cells which belong to the outer ground-tissue, close below the epidermis, begin to divide. The newly-formed cells elongate at right angles to the axis, and therefore cause small protuberances from the young stem.

The young epidermal cells which cover in these projections also begin to multiply, and thus continue to form a protective layer to the inner tissues. These begin to become differentiated by the appearance of various strands of tissue, which are the rudiments of vascular bundles.

This complex of cells which bulges out from the axis is the first stage of development of a leaf, which takes its origin from certain divisions of the young cortex, and may therefore rightly be looked upon as an expansion of cortical tissue.

The young leaf very soon ceases to grow at its apex, and

its zone of growth becomes limited to the basal portion. The apex therefore represents the oldest portion of a leaf; only in the fronds of ferns growth continues to take place at the apex, and sometimes for several years (*e.g.*, in *Gleichenia*). If a leaf is to become pinnate, lateral projections begin to make their appearance on the young leaf, beginning at the base and progressing towards the apex.

The further development of the leaf differs according to the species and the locality, the only principle of construction observed in all cases being, as we have already seen, the effort to produce a great expansion of the assimilating tissue.

If this assimilating tissue is in any danger of being dried up or damaged by too intense an illumination, nature protects it by various contrivances. Thus the *Helichrysæ* of the Cape are covered with a tremendous development of felted hairs; the epidermal cells of *Agave americana* have an enormously thickened cuticle, still further protected by a layer of wax; while *Ficus elastica* has, as we have seen, a triple epidermis, one layer of which has the function of storing water. In some cases the leaf is reduced to the form of a needle (*Melaleuca, Metrosideros*); in other cases the leaf assumes a vertical position, so that the sun illuminates the leaf edgeways (*Eucalyptus*). In this case a pallisade parenchyma is formed on both sides of the leaf. Nature can also reduce the number of the stomata according to circumstances. In many horizontal and tough leaves the shining upper surface of the leaf is devoid of stomata; in other cases they are placed in protected pits. Floating leaves (*Nymphaea*) have stomata only on their upper surface; the leaves of plants which always have a large supply of water have very large guard cells, &c.

Whatever may be the arrangement of the assimilatory tissues within the leaf, it always makes use of the first rays of light which it absorbs to develop its real working organs, the chlorophyll granules. The rudiments of these are present in the protoplasm of every newly-formed cell, and have entered it from the mother cell in the form of a small protoplasmic structure,[1] which under the influence of the light begins to form the green chlorophyll, and causes the young leaf to become green.

[1] Trophoplast.

The fully-formed chlorophyll bodies grow, and become biscuit or dumb-bell shaped. The narrow portion of the grain is almost colourless, and by becoming entirely constricted the granules increase in number. In this way the leaf attains its green colouration, which is so characteristic of the fully developed organ. By this time the cells will have attained their full size, and the intercellular spaces will have made their appearance. Even these must have definite positions to be of use to the plant. In the case of the pallisade cells they occur between the lateral walls, and thus the newly-formed substances cannot pass laterally into other pallisade cells, but must pass out at the bottom of the pallisade cell into the spongy parenchyma, or into the bundle sheath.

§ 23. **What substances does the leaf chiefly form?**

The first product of assimilation which can be detected in the chlorophyll body is the starch grain, which is the chief representative of the so-called **carbo-hydrates**. This group of substances, to which sugar, cellulose (*cell-wall substance*), inulin, dextrine, and gums belong, consists of the elements carbon, hydrogen, and oxygen, the last two substances being combined in the same proportions as they exist in water (H_2O). The starch occurs in the form of curiously stratified grains, which can be coloured blue, in some cases red, with iodine. Starch may be looked upon as the solid equivalent of the liquid carbohydrates which pass from cell to cell in the form of sugar. When the light shines on the leaf, its transpiration is greatly increased, and its cells lose some of their water; then we may imagine the starchy substance to be forced out of the thickened (*concentrated*) cell-sap, and the latter will become more liquid. The starchy substance is in most cases not entirely pure, but is mixed with amylodextrin, and this causes the stratification of the grain. The more this substance is present in the starch the redder will it stain with iodine. As soon as darkness sets in, the ferments become more active, and transform the starch into dextrin. This latter is converted into sugar, which passes from cell to cell through the cell-wall, and from the leaf into the stem, where it is reconverted into starch, and remains as reserve material for future use. In some plants the starch is replaced by fatty oils.

Ferments are substances, probably nitrogenous and of an albuminous nature, which have the power of causing the breaking up of other substances. The most common of the starch-splitting ferments (all of which are chemically little understood) is **diastase**, which causes the transformation of starch into sugar in the seeds, tubers, and stems at the commencement of the vegetative period. The sugar itself often undergoes a change by means of a ferment formed by fungi (*e.g.*, by the yeast plant), and termed **invertin**. The crystallisable **cane-sugar** (*saccharose*), which is obtained from the sugar-cane or the beetroot, is not able to be fermented, but is converted by the action of invertin (also by acids) into **grape-sugar** (*dextrose*, *glycose*) and into **lævulose**, both of which can undergo fermentation. A mixture of these two substances (inverted sugar) is of very frequent occurrence in vegetable tissues. The solution of meat, in the case of the so-called *carnivorous plants*, is brought about by "**pepsin**," which in the presence of acids (formic acid, citric acid) is a very active ferment, and can digest albuminous substances. Whether this vegetable pepsin is identical with the peptic ferment found in the stomachs of animals, which becomes active in the presence of hydrochloric acid, remains to be finally settled.

In the formation of gums, too, a ferment has been discovered. Gums often occur as transformations of cell-wall substance. The gum arabic (*arabin*) is of very frequent occurrence, and is looked upon as a combination of a carbo-hydrate with lime. In plum and cherry trees a gum, **cerasin**, is formed in all the tissues, and is looked upon as a combination of the carbo-hydrate metagummic acid and lime. **Basorin**, or gum tragacanth, is considered to be a pure carbo-hydrate formed in the genus Astragalus. Together with gums and acids we often find substances consisting also of carbon, hydrogen, and oxygen, which form gelatinous masses on boiling, and are termed **pectin substances.** They occur largely in the fruits of *Ceratonia siliqua* (St. John's bread), in tamarinds, and in the pips of apples, quinces, &c.

Glucosides represent a very largely distributed group of

substances in the vegetable kingdom, which are split up by acids and ferments into sugar and an indifferent substance. An example of this group is **amygdalin**, a nitrogenous form which occurs in the stone of various fruits (cherries, almonds, &c.), which have the flavour of bitter almonds, due to oil of almonds. This oil, as well as prussic acid and sugar, are derived from the amygdalin by the action of the ferment called **emulsin.** Another well-known substance is the **oil of mustard**; it is formed in the seeds of Cruciferæ from sulphur-containing myric acid, a glucoside, which is split up by a ferment (*myrosin*) into oil of mustard, sugar, and potassium sulphate.

Very largely distributed in the vegetable kingdom, especially in young organs, are the tannin compounds, which are chemically not very well defined. Of **tannin** itself we know that it belongs to the glucosides. Allied to them are certain colouring matters which are either dissolved in the cell sap or occur in isolated drops. Other colouring matters [1] are also devoid of nitrogen, soluble in alkalies and alcohol, but insoluble in water. They occur only in the cell-walls, and are allied to the **resins** which are rich in carbon, poor in oxygen, and devoid of nitrogen, and which occur especially in the bud scales and in woody tissues. Pure hydrocarbons occur in the form of etherial oils, which in most cases give the plants their characteristic odours. As an example we might mention the oil of camphor, which yields camphor by oxidisation.

The albuminous substances (*proteids*) are much less known because they are less easily separated one from another, and occur in numerous modification and transition stages. Chemically we may distinguish, just as in the case of animal albumen, three chief groups, the **albumins**, the **caseins**, and the **fibrins.** Albumin is the name given to the albuminous substances which are soluble in the cell sap, but which coagulate at a temperature of from 60° to 70° C. Caseins are soluble in water, do not coagulate on heating, but do so when treated with acids. Fibrins are insoluble in water but partially soluble in alcohol. The various members of these groups often differ considerably in their characters. Among the caseins, conglutin, which occurs in almonds, is much more soluble than

[1] Phlobaphenes.

legumin, which occurs in the seeds of leguminous plants. The aleuron layer of cereals contains gluten-casein and gluten-fibrin. The products of decomposition, too, of the albumens are as different as are their degrees of solubility. Thus gluten fibrin, when boiled with sulphuric acid, yields very little asparagin (a substance of great importance for the transit and formation of new albumens), while the other groups yield a large amount. Other products of decomposition are

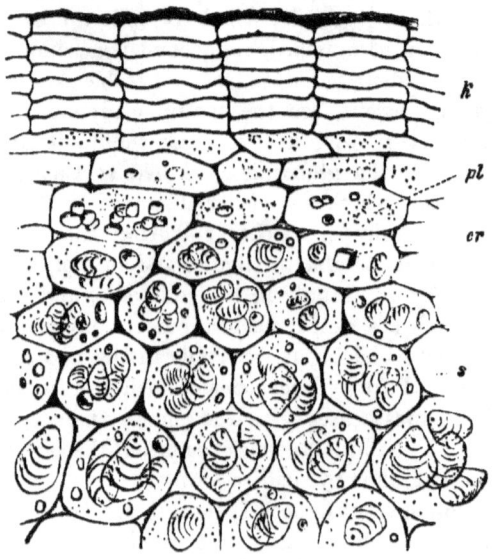

FIG. 18.—A SECTION THROUGH THE OUTER LAYERS OF A POTATO TUBER.

k, cork; *pl*, cells of cortex containing protoplasm and small starch grains; *cr*, tabloid protein crystals; *s*, starch grains (*after* TSCHIRCH).

leucin, tyrosin, glutamin, ammonia, and volatile oils. These latter cause the unpleasant odour of decomposing albumens. The albumens may also occur in a crystalline form in the cell sap; they are then distinguished from the inorganic crystals as **protein crystals.** Such crystals occur in beautiful cubical form in the end cells of the gland hairs of the potato shoots.

The protein crystals are perhaps most easily seen in sections of a potato tuber, where they occur in the cortical tissues, especially of the white varieties of potato. Indeed, the number of these protein crystals can give us some clue to the amount of starch present in the potato, for it has been found that the greater the quantity of these crystals the poorer the tuber is in starch. Fig. 18 represents a piece of a section through the cortex of a potato. The actual peel (*k*) consists of a number of cork cells which give it its toughness. Below the cork cells are the outermost cortical cells (*pl*), which are rich in albumens and poor in starch; here we shall find the most protein crystals (*cr*). The further we proceed to the centre, the larger the amount of starch contained in each cell. These starch grains (*s*) have an oval shape, and have a very characteristic and peculiar stratification.

Very often the crystalline albumens (*crystalloids*) occur within other solidly formed albumens. In the seeds of phanerogams they occur in the form of **aleurone grains,** and form the reserve albumens which are used up on germination. In these aleurone grains, which in some plants (*Pæonia, Ricinus*) are soluble in water, some albuminous compounds are mixed with phosphates of lime and magnesia, and assume a spherical form (*globoids*).

We see, therefore, that the albumens occur like the carbohydrates in form of insoluble reserve material, and in soluble transportable substance.

We must here make mention of the groups of **alkaloids,** characterised also by the fact that they all contain nitrogen. They have organic bases, containing nitrogen, carbon, hydrogen, and generally also oxygen, and are probably first formed by the protoplasm, and contained in the cell sap, but are often found in the cell-walls too. Among the alkaloids devoid of oxygen we have **trimethylamin,** which gives the peculiar odour to the ergot of rye, and some of the members of the Goosefoot family, &c.

The oxygen-containing alkaloids comprise some of our most valuable drugs, such as strychnine, quinine, morphine, atropine, which occur combined with acids in the various organs of certain plants.

§ 24. How does the leaf actually perform its assimilatory function?

The above-mentioned albumens in various proportions make up a portion of the protoplasmic substance which entirely fills the young (meristematic) cell (Fig. 19). But very soon there arise within the protoplasmic mass which fills up the cells very small quantities of liquid substance (cell sap), which collect together to form small vesicles (**vacuoles**), and these give to the protoplasm a foam-like structure. Now, as the protoplasm becomes used up during the elongation of the cells, the vacuoles increase in number, run together, and force

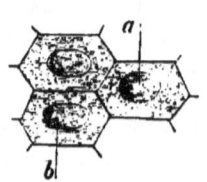

FIG. 19.—YOUNG (MERI-STEMATIC) CELLS WITH NUCLEI.
a, nucleus; *b*, nucleolus.

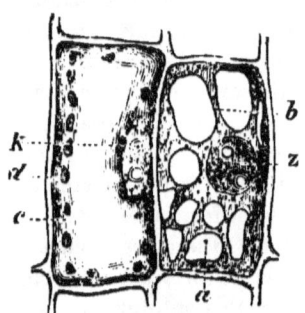

FIG. 20.—YOUNG PARENCHYMATOUS CELLS.

the nucleus (*cytoblast*) (Fig. 20, *z*) and the remainder of the protoplasm outwards towards the cell-wall. The nucleus which is visible in almost all cells is intimately connected with the life of the cell, and plays an important part in the division of the cells. It has the appearance of a dense albuminous mass, in which one or more small refringent bodies are present, which must probably be looked upon as reserve material, and which are termed **nucleoli** (Fig. 20, *k*).

With a high magnifying power the nucleus has the appearance of spongy framework, or sometimes of a number of loops of thread intimately interlaced, consisting of a denser substance, the **chromatin** substance or **nuclear thread**, the intervening spaces being filled by the nuclear sap or fluid.

The latter is a turbid, thickish liquid, which afterwards becomes clearer, and takes up certain stains less actively than the framework, which consists of the nuclear plasma (*nuclein, chromatin*).

In Fig. 21 we see in (1) a nucleus in the young mother cell of the stomate guard cell before its division; in (2) we recognise more distinctly the chromatin framework, which we see in (3) is formed of pieces of a thread. These pieces of the nuclear thread are arranged (in 4) in the median plane of the cell in which the new wall is to be formed.

They now form what is called the **equatorial plate**. The

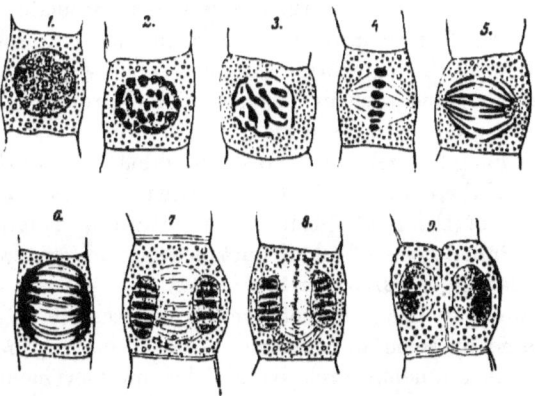

FIG. 21.—NUCLEAR DIVISION IN A GUARD CELL OF *Iris pumila* (*after* HABERLANDT).

nuclear substance now divides itself into two, and one-half proceeds to each pole of the cell (5), where it becomes arranged to form one of the daughter nuclei (6–9).

The movement of the various portions of the original nuclear thread towards the centre, and the return of the halves of these segments after division, takes place along a number of threads which are formed in the protoplasm, and are termed the **spindle threads**. These threads converge at the two poles, but separate one from the other in the middle of the cell (4–7). By the time these threads make their appearance, the former covering of the nucleus, **nuclear membrane**, has disappeared, and the threads traverse the entire cell;

the figure which they form is termed the **nuclear spindle** (4, 5, 6).

At the centre, where the nuclear plate had existed, after the division of the nuclear matter a number of very small albuminous granules (*microsomata*) make their appearance (Fig. 21, 8). They and the central portions of the spindle threads fuse together to form a plate, which forms a thin transverse plate of protoplasmic substance in the original cell. This protoplasmic plate becomes transformed into a cellulose plate, which now constitutes the cell-wall separating the two daughter-cells (9). These cells increase in size, and either divide again or pass over into permanent tissues.

By this means the daughter-cells of the stomate mother-cells are transformed into guard cells, the median wall splitting to form the aperture. In the same way, most cells increase in number, and by this means the growth of leaves, stems, and root-tips takes place.

Every daughter-cell has therefore one-half of all the constituents of the mother-cell; this explains the fact that all the characters of the mother-cell reappear in the daughter-cell, and the inheritance of all characters from one generation to another can be explained by this means.

For we may imagine that every characteristic of a cell is represented by some infinitesimal piece of matter, the atoms of which have a definite relative position and movement, which give it its character. Now, we have only to suppose that in the cell division every such particle is divided into two, just as, on a large scale, the nucleus is, and then each daughter-cell will contain all these characteristic atoms, with their definite structure and definite movement—will, therefore, have all the characters of the mother-cell. The nutrition consists in this, that the raw materials which enter the cell are attracted, now by one, now by another of the constituents of the cell, and adopt its specific movements, and are thus transformed into the same substance (assimilated). Thus every cell of a seedling contains in the innumerable microscopically still undifferentiated particles of its protoplasm as many different groups of bodies as the resting seed, but they have become active during the germination, are continually increasing in amount, and will give expression to all the characters which we recognise in the adult plant.

Though we have stated that the various groups of substances of the protoplasm cannot as yet be microscopically differentiated, we must modify the statement to exclude a few cases. For we can recognise in every young cell of a leaf, and of other tissues too, a number of small indistinct particles of a very dense protoplasmic nature, which are termed **trophoplasts**,[1] which have in many cases very important functions to perform. A number of these bodies, for instance, develop the green colouring matter, and become **chlorophyll corpuscles**, *i.e.*, the chief formative bodies of the cell; others become brightly coloured, and form the **chromoplastids** which are found in flowers and fruits; and, finally, a third group, which remains colourless, **leucoplastids**, can form starch grains out of sugars.

All these trophoplasts grow within the cell and increase in number by division. A number of them are carried into the daughter-cells during division, and we can therefore rightly say that the chlorophyll granule, for instance, in the uppermost cell of a Ricinus plant several yards in height, is derived from one of the chlorophyll bodies of the embryo.

When the trophoplast becomes transformed into a chlorophyll corpuscle, it becomes more porous and sponge-like, and in its interstices the actual colouring fluid makes its appearance. The latter consists of two substances—namely, the green chlorophyll and the yellow xanthophyll. The nitrogen-containing chlorophyll (recently shown to be devoid of iron) can be extracted from the cells by alcohol, ether, chloroform, or by etherial and fatty oils. It then has the appearance of a green solution with a blood-red fluorescence. It is a very unstable substance even within the vegetable cell, and as soon as the latter dies the acid cell-sap gives it a brownish colour. By this action acid chlorophyll or **hypochlorin** (*chlorophyllan*, *phyllo-cyanic acid*) is formed, and this is the cause of the spotting of diseased leaves. If in the cooking of green vegetables we wish to avoid such a discolouration, we need only add a little carbonate of soda, as the presence of alkalies will prevent the formation of the hypochlorin.

[1] They are also called *plastids*, *chromatophores*, *leucites*.

This substance, however, is always found in some quantity in the fresh extract of green tissues, which fact suggests that at all times a breaking down and a new formation of chlorophyll is taking place.

The less soluble xanthophyll forms in the above-mentioned liquids a yellowish non-fluorescing solution; it does not contain any nitrogen, and is less readily decomposed than chlorophyll. It makes its appearance in the cells before the green colouring matter, and it remains longer in the cells at the approach of autumn. Hence the leaves become **yellow coloured in the autumn**, and in spring the young shoots and leaves are yellowish-green.

As soon as the chlorophyll corpuscle is fully formed, it begins to assimilate, forming in most cases starch, more rarely drops of oil (in many Monocotyledons). It is certain that other substances are contained in the chlorophyll corpuscle (in some cases protein crystals and tannin masses have been observed), but a micro-chemical demonstration of the same has not as yet been possible. Speaking generally, however, of the production of organic matter in the cell, we find that the first products of assimilation are not very rich in oxygen, and that in the further changes they become more and more oxidised, until finally oxalic acid is formed, and then it returns again to carbonic acid, which is set free and given off.

So we have in the life of the leaf a marvellous cycle of changes. The leaf forms, under the influence of light, its chief formative substance (chlorophyll), and the latter starts at once its assimilation of substances, poor in oxygen, by combining the raw solution of the soil with the carbonic acid of the air. This gas, absorbed from the atmosphere, is split up in the vegetable cell, the carbon being rapidly used up, the oxygen being liberated. In this way sugar, starch, and cellulose are formed. But the untiring oxygen exerts its powers on the newly-formed substance. It unites with these new bodies and forms more highly oxidised compounds, and acids, such as oxalic acid, become more numerous. The leaf renders the otherwise injurious acid harmless by combining it with lime, and thus forming insoluble **crystals of oxalate of lime**. These have either the form of needle-shaped prisms (*raphides*), or flat lozenge-shaped tablets, or rhomboid crystals, often united in clusters. This oxalate of lime is absent from very few plants (Ferns and Horse-tails); the crystals are generally in the cell-sap, but sometimes in the cell-wall substance (Fig. 22).

In this figure (22) we see two star-shaped cells taken from a septum, which separates the air chambers of the stalk of Musa. In each of the cells, which are separated by large intercellular spaces (*i*), we see several of these prismatic crystals. Besides these we find that some cells of Musa contain the oxalate of lime in the form of raphides (*b*). While the latter occur most frequently in Monocotyledons, Dicotyledons usually have octohedral crystals, or clustered crystals of this substance; as an example of these, we have the cells (*c*) taken from the leaf-stalk of a Begonia (after Luerssen). It is more rare to find crystals of **calcium carbonate** in the tissues of plants; this generally forms an incrustation on the outer surface (many Algæ), or it enters directly into the substance of the cell-wall and renders it resistant. In some natural orders (*Urticaceæ, Acanthaceæ, Cucurbitaceæ*), the epidermis and other tissues have club-shaped or rod-shaped processes of cellulose (*cystoliths*) (Fig. 17, *c*), which have become covered with carbonate of lime, which, however, does not make its appearance if the plants are grown without lime or without light.

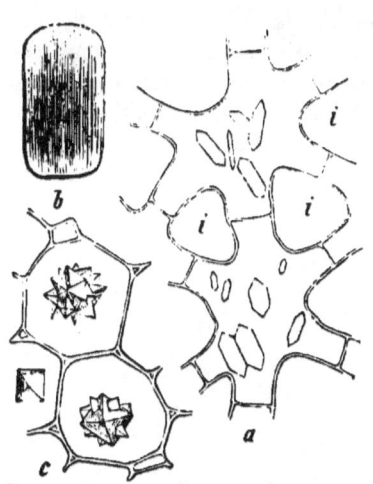

FIG. 22.—DIFFERENT FORMS OF CRYSTALS OF OXALATE OF LIME.

Only in very rare cases does the lime occur in plants as gypsum **(sulphate of lime)** or as **phosphate of lime.** This latter compound sometimes makes its appearance when some tissues which are very rich in protoplasm (potato-peels) begin to decay; generally speaking, indeed, the occurrence of compounds of lime indicates increase of age. It is only when the cells become old that the cell-walls become strengthened by salts of lime, and become less permeable; and it is then also that crystals of oxalate of lime are deposited.

Thus we find yellow autumn leaves rich in lime but poor in plastic substances. They have served their purpose. The yellow colour is due to the dissolution of the chlorophyll, the xanthophyll still remaining in the tissues. In the falling leaves the protoplasmic framework of the chlorophyll corpuscle is also absorbed, and the nitrogenous compounds make their way back into the stem. After the fall of the leaves, decomposition alone takes place within their tissues, and among the products of decomposition we have, in the first place, carbonic acid, *i.e.*, the substance which it has been the function of the leaf to absorb and to transform into organic matter.

CHAPTER VII

THE TREATMENT OF THE SHOOT

§ 25. Why must the shoots of our cultivated plants be pruned?

In all our horticultural efforts we desire a well-proportioned plant. Furthermore, we seek to increase the economic value of the plants by increasing the quality or quantity of its crop, or by decreasing the time necessary for its production.

Lastly, we have to take into consideration the portability of the plants.

Leaving out of consideration the short-lived herbaceous plants, we find that most plants which are left to themselves suffer considerably from accidental injuries (by storms or insects, &c.), which spoil their appearance. Then the fashion of the day often prescribes forms which nature would never or rarely wish to produce. We need only refer to the artificial forms of French trees, to the globular, pyramidal, or columnar forms of flowering shrubs in pots and in the open. To produce such forms, or to repair the damages to the crown of a tree, it is necessary to regulate its growth by taking away old or adding new branches.

To increase the productiveness, too, it is necessary to adopt special methods of treatment. Even when a tree is grown for the sake of its wood, it does not, when left to itself, always produce the form most desired. In growing trees for masts or tall poles, for long boards or beams, we cannot entirely rely upon growing them in closely-packed plantations. Any gap in the plantation is taken advantage of by the trees to produce strong branches, which must be removed by the forester. If with the French method of arboriculture a fruit-bearing crown is to be confined to a very limited space, the natural production of branches must be reduced, and the development of the normally quiescent buds close to the main axis must be stimulated. Similarly, if time is to be saved in the cultivation of plants, they are not allowed to retain the long shoots, but these are removed so as to cause the earlier development of flower-producing buds.

A tree or shrub left to itself often produces a crown of a disadvantageous form, by producing too many branches, so that the crown is too thick at the centre and consequently barren. Here artificial means must restore the air and light, and therefore also productiveness, to the centre. Then, again, faults take place in the growth by the production of watery shoots, which reduce productiveness of the tree (or entirely stop it for a season) if rational measures are not adopted to cure this defect.

Lastly, we have already pointed out, in dealing with the root, that plants which are to be transported must be repeatedly transplanted. Every transplantation ensures a more or less complete root-pruning. In doing this, we must remember the principle that a plant deprived of some of its roots must have its shoots correspondingly diminished. Many more examples might be brought forward to point out that the horticulturist is very rarely able to let the development of the shoots take its own course. The knife must be almost always brought into requisition.

§ 26. What is the least injurious form of a cut?

This question can be answered very briefly. **The youngest possible shoots should be pruned.** This rule is based upon the fact that the youngest shoots heal most rapidly and most completely. The danger of a wound lies in the effects which may follow upon it, and of these the setting in of decay is the most to be feared. This decay can take place without the interference of fungi, though generally these organisms play a part in the decay which sets in. The germination of fungal spores on the cut surface is brought about and favoured if the cut surface retains the necessary moisture for some time.

We must therefore prevent the accumulation of moisture as much as possible. This consideration settles at once the best form of cut and the proper time for performing it. The form of cut to be chosen must be such that the atmospheric moisture and the water which exudes from the branch are prevented from collecting on the cut. If we cut off vertical branches horizontally, the wound will be a horizontal one. But every cut branch forms, owing to the greater contraction of the pith, a cup-shaped depression at its cut end, and this depression retaining the rain-water, dew, or the sap due to the bleeding of the branch for a considerable time, will enable the germination of fungal spores to take place, with the result that the fungal *hyphæ* will destroy the tissues. A wound with an oblique position will therefore be less harmful, and all transverse cuts should be diagonal.

The precautions we have to take to avoid the collecting of moisture will also indicate that we should not choose as a season for pruning the time when plants " bleed," that is, force out much water from their tissues. Although the wounds will be found to be provided with means of preventing the giving off of water, we often see old wounds giving off a considerable amount of water. The means of protection are of two kinds. In the parenchymatous tissue a layer of separation is formed extending from the cork right across the cortex. The vessels become closed up by plugs of a peculiarly resistant gum-like mass, or by the bulging out into the vessels of the surrounding parenchymatous cells. These vascular protrusions are forced through the pits until they meet each other, and so close the vessel (*thylloses*). All these protective formations, however, arise more rapidly in the younger tissues and in those which contain a large supply of plastic substance. If we cut off a branch in a green herbaceous condition, we notice that the shoot dies down as far as the next bud, and here a layer of cork will be formed separating the dead from the living tissues. In stronger shoots, in which the wood cylinder is better developed, a swelling will often be noticed immediately below the dead tissues. Such a swelling occurs in many kinds of fruits, and is no disturbing sign; it is formed by an enormous increase of parenchymatous tissue in proximity to the hard bast fibres, which often die away for a considerable distance in the living tissue and become enveloped by a layer of cork.

A cut is least dangerous when inflicted upon a leafy shoot, first of all, because, owing to the transpiration of the leaves, the shoot will not contain sufficient water to cause an accumulation of it on the cut surface; and secondly, because the leaves will at once supply sufficient food material for the healing of the wound. If ripened wood has to be cut, the early autumn or late winter should be selected for the purpose. Early in the autumn the branches have still sufficient activity to form the closing layers, and thus to protect the shoots from the wet of the autumn and the cold of the winter. If the pruning takes place late in the winter, no injury from frost need be feared for the ripened wood, and as the tree will soon enter upon new activity, the healing of the wound will take place at an early period.

Great care must be taken in the pruning of trees bearing stone-fruits (cherries, plums, &c.), and in the South also in the case of orange-trees; in fact, in all cases where the plants are liable to the exudation of gum. If the cuts are inflicted at a time when a considerable formation of new tissues is going on, gum is sure to flow sooner or later from near the cut. A very easy way of demonstrating this fact is to make every month during the winter an incision into a cherry or plum-tree, reaching down to the wood. It will be found in the summer that, from all incisions which were made in spring, gum will be exuding. If one is forced in the case of other trees to prune during the bleeding season, cloudy and cool days should be chosen, as the pressure of the sap is least on such days. Prune as early as possible trees producing strong shoots, which, as a rule, are rich in water, and therefore would suffer most from an excessive loss of sap. The consequences of such a loss would be a retarded period of flowering, irregular ripening of fruits, and an insufficient ripening of the wood.

§ 27. How does summer-pruning differ in its effects from winter-pruning?

If the term winter-pruning is given to any removal of shoots during the resting period of a woody plant, we may say generally that winter-pruning is strengthening, while summer-pruning is weakening.

If any portion of the shoot system is taken away after it has passed through one summer, the structure and activity of the root system—that is, its power of absorption and of forcing up water—is such that it can nourish all the branches. At the beginning of the next period of activity, by cutting away some branches the water-consuming area is diminished. The same amount of pressure has therefore a reduced field of action, and consequently the effect on the remaining branches must be increased.

By pruning in the summer we remove soft shoots with only recently developed leaves. The latter have yet their chief work to perform. For at the commencement they are developed at the cost of the reserve material which is stored up in the branch; then comes a period at which the young leaf requires all the substance it assimilates from without for its own growth, and only after its full development does it begin to work for the benefit of the branch. If, therefore, a soft shoot is taken away, the older portions of the branch are robbed of the materials which were used in the unfolding of the leaves, without receiving anything in return from the leaves they have developed. This causes, therefore, a loss to the general economy of the plant; but, with the increased productiveness of our cultivated plants, such a slight weakening may be overlooked, if any other special advantage is gained.

The taking away of tips of young shoots, which is called **pinching**, is admissible in trees which are unproductive on account of the excessive development of shoots. If we take away the tip of a long shoot, we deprive it of the part which makes the greatest demand upon its nutritive powers; consequently the pressure is increased in the lateral buds, the lowest of which is increased in size while the upper ones will soon grow out to new shoots. The cells of the lower buds which are thus enlarged will store up a larger quantity of reserve material, and this must be present if flower-buds are to be formed.

The greatest success will attend this process if the pinching takes place just at the period when the buds have still sufficient time to swell up and become stored with food material, but when the supply of water begins to diminish, so that the upper buds do not grow out into long laterals.

In spring, and during the production of the so-called second shoots in June and July, the general upward current of water presents a strong difficulty. Any "pinching back" at this period calls forth almost immediately the expansion of the uppermost buds of the remaining branch. The younger the axis is, the more readily will the buds grow out into shoots if the supply of water is increased; for the cells have only shortly passed out of the stage in which their power of reproduction was greatest, and they easily reassume that condition. Thus in many trees (Peach, Ash, Acacia), in years of exceptional activity, the recently-formed buds of that year's shoots will grow out (*proleptic shoots*).

To prevent disappointment, we state emphatically, as the practice is very common, that no fixed rule can be laid down for the commencement of summer-pruning. Trees may even be pinched to death. The favourable time for this operation depends upon the climate, the soil, the variety, and even upon the individual characteristics of the plant. The cultivator must himself judge whether the shoots have reached such a stage of maturity that an elongation of the uppermost buds will not take place.

If the pruning in August has not resulted in a sufficient swelling up of the buds, in districts which have a long autumn, an October or autumn-pruning may be resorted to. In so doing, the tree may be cut back still farther, and the water supply be restricted to a smaller number of buds without fear of causing any of them to elongate. But here, too, in each individual case only can it be ascertained by experiment how far a tree may be cut back so as to produce the desired effect, namely, the production in the following year of a large number of "short shoots." In such shoots the amount of food matter assimilated by a leaf spreads itself over a much smaller stem area, and the shoot is therefore richly provided with reserve material, which is an essential condition for the formation of flowering buds.

If the pruning has been too close (which is sometimes

purposely done in varieties producing weak shoots), the pressure of sap may be so considerable in the following spring that all buds grow out into long woody shoots. In this case the tree would be **pruned for the production of wood**.

§ 28. What is the effect of the different methods of pruning ?

It is not the function of this book to discuss the methods by which a tree is forced to assume a certain shape; that is a pure technicality of arboriculture. We must, however, touch upon the consequences which the use of the pruning-knife entail to the various parts of the tree. The method of obtaining trees of definite shapes depends upon a succession of prunings carried out with a very definite aim. We need not, however, conceal our opinion that this endeavour to force a tree to assume a shape which is contrary to the natural habit of branching has no horticultural value, and only results in the production of a curiosity. This method of culture condemns itself, because the best-formed trees usually produce the least fruit. The advantages which the so-called French method of culture is supposed to have, namely, the faculty of making most use of the smallest available space and the production of especially good fruits, are reached with greater certainty if the tree is allowed more liberty to develop its natural shape. This is the case in the cultivation of dwarf varieties. The purpose of these **dwarf trees** is also to produce the most beautiful fruits in a small space. By cultivating a dwarfed stem, the crown of the tree is brought low enough for a more careful inspection and treatment, and in both cases the production of large fruits is attained by the reduction of the total number.

One of the chief methods of treatment of fruit-trees, especially of pears and apples, is to cut a branch off 4 or 5 inches above the bud of which the development is desired. Supposing this bud is to replace the main shoot as leader, all the buds above it are removed, and the eyeless upper portion of the shoot is made use of for fastening the tree to the wall. Probably in this case the food material contained in this upper portion sinks down into the lower portion when the period of vegetation commences, and can be made use of by the elongating bud. After this has grown out, the useless portion of the shoot above it is removed.

If a branch is to be completely removed, however, we must decide whether the basal dilatation at its point of insertion is to be removed as well. This basal dilatation enlarges as the shoot grows older, for that portion which is inserted in the main stem also increases by secondary thickening. The shoot which has sprung from a lateral bud was clothed at the commencement of its base by bud-scales, which soon fell away on the elongation of the shoot. But these bud-scales subtended very small and rudimentary buds, which generally remain dormant (dormant buds); now as the base of the branch enlarges, they become covered in by the cortex of the basal dilatation, and, though still present, are invisible from the outside. If, however, the branch is cut away above this enlargement, the two strongest of the dormant buds generally grow out and form two weakly shoots.

It is therefore only advisable to remove the shoot in this way if a too luxuriant branch is to be replaced by a weaker one. In other cases, it is best to remove this basal dilatation of the shoot as well, and thus get rid of the dormant buds.

This, however, is scarcely possible in the case of older branches, and we therefore often see in such trees as poplars, limes, and willows, where a branch has been sawn off, numerous small shoots making their appearance. If they are cut away, their basal shoots grow out until the number of shoots may become so excessive that they are killed by mutual pressure.

In the case of cherries and other stone-fruited trees, a different treatment is needed from that applied to pears or apples. In the case of the first-mentioned trees, the branches which have once borne flowers will not do so again, but remain bare. The formation of flowering buds, therefore, progresses gradually towards the ends of the branches, while in pears and apples fruiting spurs may bear flowers year after year, the spurs increasing continually in thickness.

In some cases, indeed, this thickening of the fruiting spurs becomes excessive; and the branch differs anatomically, too, from the normal condition. The wood cylinder is only slightly developed, but the cortex and pith are excessively enlarged, the cortex of the stem being often continuous with the massive cortex of the fruit-stalk. This form of branch is very sensitive to frosts. If it becomes necessary, in the case of such trees, to produce some long shoots, it is advisable to prune away the deformed spurs to their base, and to stimulate the dormant buds to further development.

In the case of cherries, to prevent long sterile branches, it is necessary to constantly renew the production of long shoots. This can be done by pruning back the shoot which has just borne fruit to about four basal buds. Two of these will in general grow out next year, and the stronger of the two will be retained as the fruiting branch. The same process will be repeated with this branch.

The effect of pruning, therefore, is to stimulate a number of buds to grow out which would otherwise have remained dormant. The uppermost buds of each shoot are the most ready to develop and produce the strongest lateral shoots; towards the base of the shoot the buds become more and more difficult to stimulate.

§ 29. How may pruning be used to regulate the natural development of the tree?

Among the large number of varieties of our fruit-trees and decorative plants, there are many in which the branches are not produced in a way that satisfies our wishes. Thus there exist varieties in which the branches are always bent, others in which the woody shoots are short and few in number, but which very readily produce flowering buds; others, on the contrary, are inclined to run to wood, but only produce few fruit-buds.

Similar tendencies may be produced by unfavourable situations. Pruning should in such cases be used to overcome such defects. All our efforts then will be in the direction either of producing more leafy shoots or in promoting the formation of fruiting buds. In the first case we **prune for wood**, in the latter case we **prune for fruit**.

If we find that a tree has the tendency to produce every year a number of shoots with an excessive number of fruiting buds, so that the tree is in danger of exhausting itself and not producing a proper crown, the tree must be pruned for wood. We know that a rich and continuous supply of water increases the vegetative activity of a shoot, the leaves become larger and the internodes longer. The supply of water to the various buds can, however, be relatively increased by decreasing the number of buds. The number of eyes will therefore have to be diminished by pruning the branches, and the stronger the shoots are to be next year, the shorter the branches must be cut. Pruning for wood is therefore merely a cutting away of branches, leaving only a short piece of each shoot.

Pruning for fruit, on the other hand, becomes necessary when a tree produces long unbranching shoots of a vigorous character, but which show no flowering buds. This production of woody shoots can only be stopped by cutting away the tips of the branches, leaving the greater portion, or at least the half, of the shoot. The uppermost buds of the remainder of the shoot soon grow out, but are pinched off while still young, and by this means the lower buds are encouraged to swell and to form flower-bearing buds. If this method of procedure does not result in a sufficient production of fruiting buds, bending, twisting, or ringing the shoots may be tried. We shall refer to these special means a little farther on.

The same variety will have to be treated differently in different localities and also in different climates. Colder climates have the same effect as damp ones of causing a production of wood, and necessitate therefore a similar treatment. Warm, dry, and sunny localities promote the formation of short shoots, rich in food material, and are therefore favourable to the production of flowering buds; the trees in such localities must therefore be well pruned in, so as to stimulate the growth of strong branches, which are necessary for the full development of the crown.

Pruning should also be used to correct the faults of over-manuring. If, for instance, a tree has been stimulated by an excessive amount of nitrates or potash in the manure to produce too large a number of long shoots, such over-production of wood can be stayed by close pruning. By doing so the weak buds at the base of the shoot are caused to swell and to develop more strongly than would normally be the case. By this means strong shoots will later on be formed in the neighbourhood of the main branches.

If the general constitution of the tree is weak, it is essential to create as large an amount as possible of assimilating leaf surface. In such a case, therefore, it is advisable to prune the shoots only slightly, giving at the same time sufficient water to the roots.

Varieties which flower sparingly should not be pruned in closely, but this may take place in the case of plants which produce a large number of flowering buds.

The same close pruning is applicable to varieties in which the growth of the apical buds predominates and which produce very few lateral branches, so that the crown when left to itself consists of long unbranching shoots. If the varieties branch freely they should be only slightly pruned.

The character of any variety is, however, greatly changed by the stock on which it may be grafted. The so-called dwarf-stocks of our pears and apples (paradise-stock and quince) cause the production of shorter shoots, but, on the other hand, a greater and earlier production of flowers, whereas the wild (crab) apple and pear stocks favour the formation of strong woody shoots. Grafts on the latter must therefore only be slightly pruned, whereas the others may be closely pruned.

The age of the tree, too, requires to be taken into consideration. All young trees tend chiefly to develop vegetative shoots, *i.e.*, to run to wood, and should therefore be pruned less closely than old specimens of the same variety and in similar situations.

In the case of trees grown against a wall, especially if the artificial French shapes are the object of attainment, the pruner must pay attention to the position of the branch. In the "cordon" type the strong branches are kept horizontal. But the more horizontal a branch, the smaller the supply of water to the buds; the nearer the branches approach the vertical, the more copious is the supply of water and the larger the leaves and internodes. The shoots coming off from the horizontal branches, therefore, must be more closely pruned than those of the vertical branches. An exception to this rule must be made in the case of the watery suckers which are produced in the case of "cordons" at the bend of the leading axis. These soft shoots are caused by the unnatural horizontal position which is forced upon the main stem. On account of the bending, a certain number of buds at the bend are more plentifully supplied with water. When the branches are left in their natural position, the buds near the apex develop most strongly; but when the branches are bent to the horizontal, the passage of water is retarded, and the supply, therefore, to the buds immediately at the bend is abnormally great. This causes in the first place a greater turgidity and enlargement of the cells of these buds, and also supplies them more liberally with nutritive salts. When once such a bud has started growing, its leaves will constitute a further centre of attraction for the raw sap, and being in such a favourable position, will elongate very rapidly, becoming a strong and vigorous shoot. This, however, entails a loss to the mother-stem and literally robs it of its nutriment. Thus the new vertical shoots may often cause the death of the original horizontal axis.

If by any inattention such a "robber" has been allowed to develop, its tip should be pinched, but it should not be entirely removed at the outset. Only those which most disturb the crown should be pruned to about a third of their length, and their laterals should be nipped off as soon as they develop. No general rules applicable to all cases can be given; this only can be laid down as a guiding principle, that the development of such branches is due to the fact that the amount of water taken in by the roots is not in proper proportion to the existing working surface, which is for the time too small. If, therefore, the vertical shoot is suddenly taken away and the surface again reduced, this procedure will only result in the production of another shoot of the same nature. They should, therefore, if possible, be used for the normal development of the crown, or removed slowly after having encouraged new and more suitable shoots.

§ 30. When is pruning harmful?

It is contrary to good arboriculture to prune a tree too much, so that it is covered with cuts. In our nurseries and gardens trees are often too closely pruned, a procedure which often diminishes productiveness and causes many diseases, not infrequently indeed kills our pyramid trees.

The excessive pruning is caused as often by too frequent application of the knife as by removing too great an amount of wood.

We must give the tree an opportunity of developing a few strong woody shoots, which will only be shortened after the wood has ripened and the leaves have fallen, otherwise the tree will become weakened. **To remain in good health and continuously productive, the tree must have a sufficiently large supply of leaves, and these must remain active as long as possible,** so as to store up a large amount of food material in the stem. But if every strong shoot is pinched as soon as it exceeds the desired length, and the lateral buds are thus caused to grow out, and if they in their turn have their tips removed, the tree will not have a sufficient supply of well-developed leaves, and a large portion of the food material formed by them will be used up by the developing lateral branches. It must not be forgotten that at the commencement **every young shoot draws like a parasite upon the food matter of the older branch**; this applies as much to the consumption of water as to the stored-up food material. The need of a strong root or shoot system for a sufficient supply of foliage is often seen in the case of cordon apple-trees. An exceptionally vigorous leafy shoot will suddenly make its appearance, and if removed, will cause the production of a number of vertically growing suckers.

Pruning "to excess," however, may also be caused by the untimely removal of older branches, as was formerly often the case in the raising of standard fruit-trees. By cutting away the lateral branches at a very early period, the stem, it is true, would grow very rapidly in height, but the annual rings were only feebly formed, and the stem was often too slender to bear the weight of the crown. The modern practice is to reduce the leading shoot, and thus cause a stronger development of the lateral branches; these will be able during the course of the summer to send a sufficient supply of nutritive matter into the young stem, which will be able to use it for the development of broad rings of wood.

The tree may furthermore be injured by removing too many branches, so that the tree will not have sufficient centres for the use of the sap raised by the roots. If the branches were removed bit by bit, growth would not be interrupted. It is the sudden removal of the branches which acts injuriously. Too many of the readily elongating buds near the end of the shoots have been cut away, and this may cause excrescences of the cortex and cambium, which may become centres of future disease. This is the case in a disease to which the gooseberry is very liable. If, in the winter grafting of *Ribes Grossularia* on *Ribes aureum*, the stock is too much cleared of its lateral branches, the cortex below the insertion of the scion will often form swellings, due to tube-like elongation of its cells, which may ultimately rupture the bark and prevent the graft from growing on. Similar swellings, which do not, however, rupture the bark, are formed on the spurs of soft varieties of pear; their formation is preceded here too by a removal of all the buds of the spur.

If the cambium layer suffers from an excessive supply of water owing to the removal of the lateral branches, the newly-formed annual ring may become interrupted by transverse bands of short soft parenchymatous cells in place of the long and hard prosenchyma, and these bands may extend a good way from the surface exposed by the cut. These bands of soft wood parenchyma loosen the annual ring, and may be a great source of danger if any signs of decay show themselves at the cut end. For this softer tissue is very liable to decay, and easily attacked by fungal growths. Trees which are apparently quite healthy will often show, when sawn across, brown markings in the white wood, very often a complete brown ring of decayed tissue. In splitting the stem longitudinally, the more central axis will split away from the outer hollow cylinder of wood just at this brown ring. Such stems are of course useless for cutting into boards or planks. Lastly, we come to a point which in practice is very rarely sufficiently appreciated, namely, the sawing off of thick branches. There are, of course, cases in which this procedure cannot be avoided, as, for instance, in the removal of the crown for purposes of grafting or after damage by wind or snow; but there are many occasions when this always somewhat dangerous experiment might be, but is not avoided. This is especially the case in putting up telegraph wires along wooded roads and lanes and in thinning woods.

The damages that ensue are of two kinds. First of all, extensive wounds are made, which take years before they are completely covered up, and during that time they give access to wood-destroying Fungi. Secondly, the equilibrium of the crown is disturbed in such a way that a sudden production of numerous soft shoots takes place inside the crown, which becomes choked and loses many of the horizontal fruit-bearing branches.

Some, who attach little importance to the removal of large branches, argue that in woods such branches are often removed to obtain straight trunks, and often decay away naturally on account of the want of light, and in neither case does the tree seem to suffer.

But in both cases the argument falls to the ground. For in the practice of forestry the removal of branches for the benefit of the main stem often induces the decay of the tree, and in the case of branches breaking off for want of light, the branch is dead before separation takes place, and the cells and vessels are sealed up with plugs of gum, resin, and by thylloses, and are rendered unfit for fungal growths.

Of course the trees vary very much with regard to their powers of resisting any such injury. For while in Conifers the wound is covered up with resin and the stump of the branch becomes soaked with resinous oils, which defy fungal attacks, this is not the case with most Dicotyledons. In our fruit-trees, a wound of two or three inches in diameter, which becomes covered in in from six to eight years, always causes brown spots of injured tissues in the stem. With larger wounds this is still more the case. Decay will then almost invariably set in in the branch, and subsequently in the stem. Even artificial covering of the wound is not always sufficient to prevent this, as the wood which is laid bare will produce large cracks when it contracts in seasoning, and so offer a point of attack for Fungi.

§ 31. In what way can the natural process of healing be accelerated?

First of all, we must see how the natural process of healing is effected. The age of the injured organ is of the foremost importance, as the process of healing takes place more rapidly the younger the injured tissues are. If a root, for instance, is deprived only of its meristematic apex, the entire wounded surface will begin to form new cells, and the root-tip will be entirely renovated. If, however, the axis is cut below the apex, where the tissues have already become differentiated and the woody cylinder has made its appearance, we notice not only a retarding of the process of healing, but we observe also that all the tissues are not able to take part in the closing of the wound. The outermost, that is, the oldest regions of the cortex, the woody cylinder, and the pith too, are not able to enter again into cell division.

In a branch of one year's growth the formation of cells which will gradually cover in the wound is restricted to the cambium layer. It is not, it is true, only the cambium layer of the anatomists, *i.e.*, the single layer of meristematic cells, but also the cells adjoining it on either side, which represent the youngest cells of the bast and of the wood; still it is only a comparatively narrow ring of cells just outside the wood which possesses the power of forming the healing tissues. In some plants (the lime), under favourable circumstances the pith too of the first year's twigs can take part with its outermost layers in the formation of new cells, but usually this faculty is not of much consequence. The tissue formed by the cambium and spreading over the cut surface consists at first of delicate meristematic cells (*callus*); later on, the cells become differentiated in such a way that those which are placed externally form bast, while the more internal ones which cover the wood form short lignified wood cells. Thus the meristematic callus forms gradually the closing layers of wood which cover in the cut surface, and which may be looked upon as a continuation of the tissues which formed the branch. The chief feature which we have to notice is that the cambium of the branch is continuous with the cambium which forms the closing layers. Upon the activity of this cambium depends the annual advance and increase in thickness of the layers of wood which ultimately cover the cut like a cap (Figs. 23, 24, 25).

The function of the callus and closing layers will be readily understood from the adjoining figures. Fig. 23 shows the formation of callus on the surface of a woody tissue which has been laid bare; this will take place under favourable conditions in all stems in which the bark and bast has been peeled off during the period of special activity of the cambium. In this case the youngest layers of cells on the surface of the wood will enlarge to form rows of delicate cells (*c*) which are united into a continuous layer. This tissue is formed not only by the cells (*m*) which form the continuation of the medullary rays, but also by those (*l*) which adjoin the wood fibres. In comparison with the fully developed elements of the wood (*h*) the cells of the newly-formed tissue are exceedingly thin walled and full of protoplasm; but they chiefly differ from the former by constantly growing in length and cutting off cells at their apex. Later on this active growth in length ceases, and the last-formed cells and those immediately below them become divided by closely set parallel walls, which become corky, and thus form a layer of corky cells (*k*) which protect the delicate tissue. This concludes the growth of this tissue as callus. Then within the mass (*c*) a new meristematic

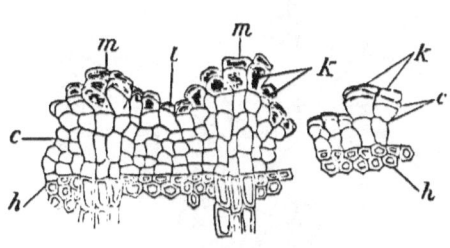

FIG. 23.
THE FORMATION OF CALLUS ON THE EXPOSED WOOD SURFACE OF *Tilia europaea*.
(Explanation in the letterpress.)

layer makes its appearance; this is a layer of constantly dividing cells which runs parallel to the old wood (*h*), and this layer behaves like a true cambium layer, forming new elements of wood towards *h*, and new bast cells towards *m*, *l*, *k*. In this way the surface of the stem which had been laid bare covers itself afresh with a new bast.

Fig. 24 represents another method of callus formation, which we shall refer to again in speaking of grafting. Here the callus is formed entirely from the cells of the pith. The figure represents one side of the innermost angle of a wedge-shaped cut which has been made into the wild stock for the purpose of inserting the scion. The cut has reached down to the pith (*m*), and has cut the woody cylinder (*h*) along the line *sch*. The opposite side of the wedge-shaped incision has been left out of the figure. In consequence of this cut a growth of rather rare occurrence can be noticed—namely, the growing out of the cells *mk* at the juncture of wood and pith. They

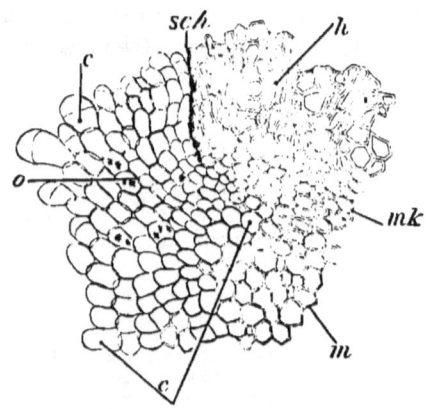

FIG. 24.—FORMATION OF CALLUS IN THE PITH OF A ONE-YEAR-OLD TWIG OF *Tilia europæa*.

have not only grown out, but have divided to form a considerable amount of callus, reaching from *c* to *c*. This callus mass grows into the crack which exists between the stock and the scion, and aids the growing together of the two, which is, however, chiefly caused by a similar but more abundant growth of callus from the cambium layer. Crystals of oxalate of lime (*o*) are seen at intervals in the mass of callus.

In Fig. 25 we see the method of healing and the formation of callus from the bast and cortex. The cortex and bast were cut through in the direction of the arrow, the cut reaching down to the old wood (*ah*). The reader must remember that he is looking at a transverse section of the branch. At the time the injury took place, the cambium layer (*camb*) of the branch was in contact with the old wood (*ah*); on the outside of the cambium was the old bast (*r*). After the cut had been made, those cells of the bast and cortex which were still capable of division, as well as the cambium cells, grew out into the gap and formed the callus layers (*c* to *c*).

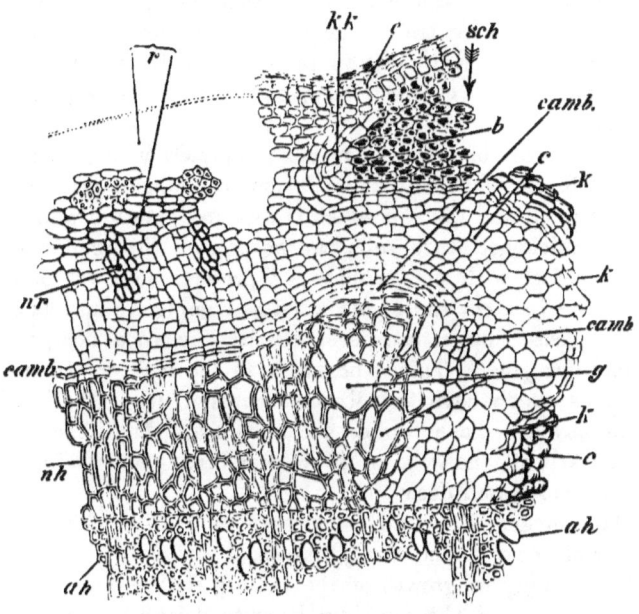

FIG. 25.—THE HEALING OF A CUT IN THE OUTER LAYERS OF A BRANCH OF *Tilia europæa*.

Very soon the callus formed a protective cork layer (*k* and *k*), and the latter was continued backwards to the cork layer (*kk*), which had been produced by the cortex cells immediately below the original cork, and which thus separated the injured, and consequently dying, bast fibres from the living tissues. In this way a continuous protective layer was formed round the injured tissues, and the cambium layer (*camb*) started forming new wood on the inside. This new wood (*nh*) was of a looser nature, and contained more and larger vessels (*g*) than the old wood (*ah*), and the latter became still more separated from the old bast (*r*) by the formation of new bast (*nr*) on the outside of the cambium. At the point where the cut was inflicted, the cambium (*camb*) bends down into the original callus and forms there too new wood and bast, which form the closing lips projecting from the healing tissues.

If we imagine the branch to be cut through transversely, we find this healing tissue starting on all sides as a ringlike outgrowth from the cambial and cortical cells, and gradually approaching the centre of the cut surface, until the margins of this circular crest meet in the centre. Here the margins fuse and form a complete cap, covering the stump of the branch.

This process of covering may be completely successful, but it must be remembered that no fusion ever takes place between the old wood and the cap which covers it. Mature and completely thickened wood cells can never give rise to new cells, nor ever fuse with new tissues. A protective growth of cells may cover in a cut end of old wood as completely and as close as you please; but it can never nourish the wood nor prevent it from decaying. This is best seen in the case of stumps on oak-trees which have thus become covered up. The healing tissue may here have formed a complete cap, but if the latter be knocked off, the wood of the former branch will be found to be discoloured and decayed.

The quicker a wound is covered in, the smaller is the danger of decay to the woody tissues.

The rapidity of growth of the healing tissues depends upon the proportion of the cambium layer to the area which has to be covered. Consequently the conditions become more unfavourable the larger the cross-section of the branch.

Furthermore, of two equally large wounds, the one in which the cambial activity is greater will be the first to heal completely. The rapidity of cell division, however, depends upon the amount of nourishment which the cambium receives, and this again is dependent upon the position and form of the wound.

The position of the wound is of importance, inasmuch as

different regions of the stem or branch are nourished in different degrees. If above the wound there are plenty of leafy shoots which are actively producing plastic material, the descending food material will enable a rapid covering in of the wound to take place.

The shape of the wound inflicted is also of some importance. If we examine, for instance, the place where a piece of wood has been cut out of a stem, we notice that the healing tissue is more profusely produced at the upper end of the injury. This is due to the fact that the plastic substances descend from the leafy portion of the tree, and the upper edge of the wound arrests this downward current and causes there an accumulation of plastic material. If the wound is proportionally very broad, a large amount of this material will accumulate, and the healing will be more rapid than if the greatest length of the wound were parallel to the long axis of the stem.

With regard to the rapidity of healing, it is interesting to notice the difference between a transverse and longitudinal cut. If a branch is cut across transversely, there remains of course no leafy tissue above it, and the cambium ring must draw all the material it needs for callus formation from the immediate vicinity. But plastic material does not travel upwards to any great distance, and hence it will be observed that comparatively small transverse wounds will not heal, or only do so very slowly, if they are far removed from a lateral bud. If you desire to cut a branch transversely, it is desirable to do so immediately above a bud, because in this region there is a larger store of food material, and also when the bud grows out into a shoot, it will send down more food material which will become available for the healing tissues.

From these considerations we may deduce the general rules, that **wounds are covered up more readily, the less numerous they are on any axis, the smaller their area, the more they depart from the form of a transverse cut, and the greater the amount of active leaves which remain above them.**

Any artificial aid in the process of healing must therefore be based on these considerations. Hence we generally prune a branch by a long oblique cut close above a bud. It is also advisable to cut a branch with a sharp knife, and in the case of thick branches, to pass over the sawn surface, at least at the margin, with a knife. If the surface is very rough, so that the injured and dying cells terminate at different heights, the atmospheric moisture remains in contact with it for a long time and in considerable amount, and offers facilities for the spores of parasitic Fungi to lodge themselves on the surface and to germinate there. In large wounds, which take many years to cover, it is necessary to protect the wounded surface from atmospheric influences. Of all the substances used for this purpose, those should be preferred which produce the same effect that is produced by natural means in the case of Conifers. But as the resinous substances used for this purpose are still too expensive for universal application, **tar** is most generally used. On account of the cracks, however, which gradually appear in an old wound which has remained open, it is advisable to **repeat the tarring.**

§ 32. **By what means can we increase the effect of pruning?**

We do not always attain by pruning the end we have in view, and this is especially the case when we seek to accelerate or augment the productiveness of our trees. In consequence of this failure, recourse is often had to operations the nature of which we will now examine.

(*a.*) *The Bending of Shoots.*

The operation which gives least irritation to the economy of the plant is the bending of shoots.

In most cases where these artificial means are employed, branches are bent from a more or less vertical towards a more horizontal position. From our preceding considerations we know the effect of this procedure. First of all, the nutritive sap becomes unevenly distributed. In their natural position the water supply is most favourable to the buds near the apex of the shoot. If the shoot is bent, however, so that the apex is at a lower level than the middle of the shoot, the buds at the highest point of the arch formed by the shoot will be the most favoured. If the whole shoot is laid horizontally, the supply of water to the whole branch is diminished, as the bend at the base of the shoot retards the current of water which is directed towards the apex.

If you examine the under surface of a greatly bent shoot at the bend, you will see that the outer tissues are compressed into considerable wrinkles. This is explained by the fact that the cortical tissues have had to be reduced within the bend to a smaller area. These outer wrinkles are, however, only an external manifestation of an internal damage which consists in the splitting away of cortex and bast from the internal wood. The space thus formed between the wood and the bast is, however, not apparent from the outside, and later on a microscopic investigation does not reveal the existence of a split within the bast, but shows a whitish hardened tissue. This is newly-formed, short-celled, and parenchymatous wood, which is rich in starch. Any considerable bending of a branch entails, therefore, at the bend a loosening of the tissues in the lower side of the branch and an increase of tension on the upper side, so that the cells on this side are stretched and consequently become narrower. By this lateral compression of the cells of the upper side the conducting tissue, both for the plastic material descending from the leaves and also for the rising sap, has its activity diminished. The kinking of the tissues on the under side produces a similar interruption of the flow of substances. Considerable bending of a shoot always diminishes the flow of sap towards the apex, and also the conduction of assimilated material towards the base of the shoot. The latter accumulate, therefore, at the upper portion of the bend; the raw sap, on the other hand, is stopped on the lower side.

Every bend acts, therefore, as a barrier to the ascending and descending currents. Above the bend the plastic material accumulates and causes a more ready development of the buds at that point into flower-buds. Below the barrier the buds are more liberally supplied with water, and develop, therefore, into strong leafy shoots.

The effect of the bending depends chiefly on the mechanical

change which the tissues undergo at the bend, and must, therefore, show itself also when branches which naturally grow in a downward direction are bent upwards.

(b.) *The Twisting of Shoots.*

The twisting of shoots acts in the same way, but much more energetically. In this case the shoot is twisted half-way round during its period of growth and while still in its normal position. By so doing the woody portion in which the twisting takes place is loosened, and probably split up in the interior. Afterwards the shoot is also bent permanently downwards at this point, so that the shoot forms a loop with the tip of the shoot pointing downwards. At the bend the under surface of the twig comes to lie above, and the wood is split up into a number of spirally twisted loops.

The region at which the twisting takes place forms a considerable swelling, in which, as is revealed by the microscope, a formation of parenchymatous tissue has filled up the splits. The cambium, too, has gradually re-formed into a continuous ring, producing new wave-like masses of wood round the broken tissues.

The breaking up of the woody tissues and the twisting the more delicate bast necessitates of course the loss of the upper portion of the branch, and all that is effected is a diminution of the upward passage of water and a supplying of the buds immediately above the bend with the assimilated material which continues to be formed in the upper leaves of the shoot.

In **layering** quinces, the shoots are often twisted about their long axis at the point at which the roots are to be formed. The plastic material passing down from the leaves accumulates at the twist, and can be used for the production of adventitious roots.

In the Caucasus it is usual to twist the fruit-stalks of the ripe grapes, so as to produce an especially good sort of wine. The explanation of this procedure lies in the fact that the supply of water to the grape is diminished, while the evaporation from the berries remains the same; consequently, the sugar-containing sap within the berries becomes more concentrated, *i.e.*, more sweet. The process of respiration, during which a portion of the organic acids is oxidised to carbonic acid, is not interfered with by twisting the stalks.

The greatest retardation of currents is produced by **breaking the branches.** If a shoot which has already formed its secondary wood is broken over the blade of a knife, a splintered wound is formed. The tip of the shoot is connected with the base by a certain amount of bast and cortex and a very small amount of wood. It dries up more rapidly the more woody the region in which the break occurred. If the shoot is soft, the wood is not splintered, and the drooping end of the shoot very soon raises its tip and begins to bend upwards, but grows very slowly indeed.

(*c.*) *Notching.*

The fruit-grower is often compelled, so as to obtain trees of a regular shape, to force a certain bud to develop without being able to alter the position or form of branch to which it belongs. In such cases, he has to resort to notching, *i.e.*, to remove a narrow sickle-shaped mass of cortex and young wood on one side of the branch immediately above the bud. By so doing, the upward current of water on this side of the branch is diminished above the bud, while the latter, being more abundantly supplied, will grow out very rapidly. If the notch is made below the bud, the vascular bundles which supply the leaf are usually severed, as they leave the woody cylinder of the stem a good way below the leaf and run in the cortex for a considerable way. The notch below the leaf stops the plastic materials from passing downwards, and they accumulate chiefly in the form of starch in all the parenchymatous cells in the neighbourhood of the bud. This storage of food material is considerable, as the woody cylinder is interrupted opposite the bud, and a parenchymatous bridge reaches from the pith right through the wood to the bud. The tissues surrounding the bud thus become laden with food material at the same time as the water supply diminishes. Such conditions are very conducive to the formation of flowering buds. If a shoot is developed, it remains short, often indeed bunchy, and is very prone to produce flowering buds.

If the notch heals up, it still remains of some importance for the next year, as the wood first formed in that place is large celled and will be able to store up starch. Still it is advisable to repeat the notching if the desired result is not obtained in the first year.

What has been said here about the bud is equally applicable to a shoot or branch. The size of the notch must, of course, depend on the size of the branch. In a one-year's shoot, the breadth of the notch may vary from one-twentieth to one-quarter of an inch, whereas in a strong branch it may rise to one-half or two-thirds of an inch. The best season for performing this operation is the late summer or beginning of autumn, as the buds will have time to gain some advantage from the altered conditions of nutrition.

(d.) Ringing.

Ringing is the term given to the removal of a narrow ring of cortex and bast during the period of greatest cambial activity, when the outer layers come away most easily from the wood. The wound dries up very soon, as the outermost layers of the young wood which is exposed (splint-wood) are very readily dried. This reduces the upward conduction of water to a slight extent, and in the same way the upward passage through the cortical cells is stopped. That such a passage from cell to cell does actually take place can be seen when strips of cortical tissue are separated from the woody cylinder, but are left attached to the branches at their upper ends. If in such cases the transpiration from the wound is prevented by surrounding the branch with a glass cylinder, it will be found that the strips die off below, but remain turgid near the upper end, and often form new wood cells on the inside. Water, therefore, with the nutritive salts it contains, moves along the cortex independently of the wood, and thus supplements the action of the latter. But besides the reduction in the passage of water due to the removal of tissues, the upward current is further weakened by evaporation from the exposed woody tissues.

In the cells above the excised ring of cortex the chief factor in the elongation of the cells, the turgidity of the latter is reduced. The elongation of the cells is lessened, and this manifests itself in the reduction of the apical growth and the diminished length of the internodes. The plastic material which comes from the upper end of the branch accumulates above the wound and causes an increase in the activity of the cambium and a greater storage of reserve material. We notice an increase in the diameter of the upper portion of the branch compared with the portion below the ring-shaped cut.

A shortening of the internodes, however, and an increase in the amount of plastic and of reserve material, is the first condition necessary for the formation of fruiting buds; thus by ringing the branches become productive at an earlier period. Branches which have been ringed assume the autumn tints at an earlier period and ripen their fruit earlier. Such a hastening of the ripening is especially desirable in damp autumns in the case of late varieties of vines, as the grapes would in such cases often not ripen at all. The success, however, of ringing these plants, especially in the case of American varieties and hybrids, will only be assured if a considerable amount of splint-wood is also removed. For the very wide vessels of the wood will conduct very considerable quantities of water to the growing points, so that the growth in length will scarcely seem to be reduced. In the case of vines this procedure is used against the dropping of the berries.

Ringing, however, must never be looked upon as one of the regular operations in arboriculture; it will always be an unnatural treatment, which should only be used in exceptional cases, as it generally entails the early death of the branch. Only in those cases where the entire branch can be sacrificed should one resort to ringing. For this operation does not only affect the portion of the branch above the wound, which easily breaks off, but the base of the shoot or branch is also endangered. This latter portion receives no plastic material from above, the cambium ring is therefore starved, and the buds, owing to their deficient nutrition, do not form flowers and fruit.

It is more advantageous to make only a very narrow circular cut on very vigorous shoots, so that the wound may become closed during one vegetative period. After the healing has been effected the upper end of the shoot can continue its growth in length, and the lower portion will again receive the necessary material for the normal increase in thickness of the wood and for the production of buds. This can be brought about with certainty if the branch is constricted with a ring of wire. The growth of wood at this point is in the first place altered by the pressing together of the wood fibres, which become laterally twisted. Very few true vessels are formed between them. Afterwards, when the wire becomes enveloped by the new growth, the cortex and bast are ruptured and a soft wood is formed consisting of short, wide parenchymatous cells. Of course, as long as the wire has not been grown over the branch is liable to break off; but in other respects no care need be taken for its further development. A deep black discolouration of the wood will be noticed in the neighbourhood of the wire, but no attention need be paid to this appearance. It is caused by the combination of tannic acid with the iron and penetrates far into the wood. It is the layers of splint-wood formed after the concrescence has taken place which form the chief conducting tissue for the upper portion of the branch. The sufficiency of the layers of splint-wood to supply the entire leaf system is apparent from the vigorous growth of hollow willows.

(c.) *Peeling the Stems.*

The most dangerous operation which is performed upon the living tree, and which is a matter of life or death to the individual, is the process of peeling off the bark. In such cases a large portion of the bark of a thick branch or stem, extending in length from six inches to three feet, is removed. It is therefore a process of ringing on a large scale, and we only adopt it as a last resource when all other methods of making a tree productive have failed.

The danger of this operation is the uncertainty of the heal-

L

ing of the large surface which is exposed by the wound, but it is by no means impossible.

The process of healing, however, is different from that of all the other operations which we have so far considered, in all of which the exposed tissues became covered in by an overgrowth from the margins of the wound. But in the case of a wound the upper margin of which may sometimes be a yard away from the lower one, it is evident that the margins will never be able to coalesce. Yet this large area, if it does heal, will heal more rapidly than that exposed by ringing, in which operation the knife removes the youngest layers of wood.

In peeling the stem, however, it is essential to leave the new wood intact, for it is from this layer that the healing tissues are formed (see Fig. 23). How very important it is not to damage these layers of splint-wood we can see from the appearance of those regions where the knife has made longitudinal incisions to enable the bark to be lifted. The operation, which can therefore only be executed when the cambial activity is very great, and when a very delicate layer of cells exists between the bast and the wood, is commenced by making one or (in the case of thick stems) several longitudinal incisions reaching down to the wood. These incisions reach from an upper circular cut down to the point where the lower cut is made. Where the circular and the longitudinal cuts meet, we begin to lift the bark and then peel it off from the underlying wood. If it comes off easily, the exposed wood is smooth and moist.

If it is smooth and free from adhering portions of fibrous bast, there is a prospect of the operation succeeding. If the surface is not clean, the healing is incomplete, or may not take place at all.

A few days will be sufficient to ascertain whether the tree will live or die. If the latter is going to happen, the wound turns grey and shows black lines. These are due to the development of young fungal colonies, which attack and destroy the woody tissues. If, however, the healing process is successfully established, the wound assumes a yellowish-green colour, and in a few weeks we can ascertain by pressing with the finger-nail upon the external tissues that a layer of soft tissue has already been formed over the exposed wood.

This layer has been formed from the youngest cells of the wood, which must have remained undamaged while the thin cambium cells were ruptured, have thickened their outer walls immediately after the operation, and have thus protected themselves against the drying effects of the atmosphere. Sometimes this protective layer is strengthened from the beginning by the collapse of the outermost layer of cells, while the cells immediately beneath begin to multiply and form the new tissue. We see the exposed medullary ray cells beginning to divide; the cells which were previously destined to be wood fibres divide again and form short cells, and even in the young vessels thylloses are formed and divide up to form new cells; in fact, all the cells constituting the new wood develop into a homogeneous callus-like layer of cells, which at first looks like a layer of cortex cells. But as these cells multiply a differentiation begins to take place. The innermost layer alone remains meristematic and forms a connection with the normal cambium of the undamaged stem above and below the wound. A new cambium layer is therefore formed over the entire exposed area, and is continuous with the normal cambium. It forms the new elements somewhat shorter at the commencement in exactly the same way as the normal cambium, and thus a new layer of bast begins to cover in the wood of the exposed area.

This account of the process of healing explains two rules which have to be observed. As the material for the formation of the new layers of cells has to come from the reserve material stored up in the wood, the most favourable time for the operation is that period of cambial activity at which a large amount of material is again becoming stored up in the wood, and that is the period **immediately preceding the second period of growth.** In the spring the cambium, it is true, is very active, and the bark can be easily removed, but the reserve substances are already transported away, and are being used up in the unfolding of the buds, and there is therefore too little material for the healing of an extensive wound.

The second rule, to which also sufficient attention has not

been paid by gardeners, is the time of the day at which the operation should be performed. It has been usual to prefer the springtime for this operation, because it seemed desirable to avoid the great heat of the summer. It was supposed that the latter would dry up the wound too rapidly, and therefore dull days were selected for this operation, or the wounds artificially shaded or protected. But these assumptions are wrong and have led to frequent failure. We have seen that the exposed area protects itself by the thickening of the outer cell-walls or by the collapse of these cells. Now both these processes are accelerated by light and heat. Long-continued moisture, on the other hand, will facilitate the attacks of parasitic Fungi. It is therefore best to **peel trees in hot summer weather** immediately before the second growth in August, and to do so **in midday heat.**

§ 33. Why do we slit the bark?

By slitting the bark we mean making longitudinal incisions, in which the knife penetrates into the region of new wood, but does not remove any of the tissues. These incisions relieve temporarily the pressure of the bark upon the wood in the regions which are cut, and prevent the trees from becoming hide-bound. The wounds thus inflicted heal very rapidly by a closing in of the two margins of the cut, and in accomplishing this a certain amount of the substances stored in the wood is used up.

This operation may therefore be undertaken either when the cortical pressure in stem or branch becomes so great as to prevent the formation of the necessary amount of new wood, or when more plastic material or sap is stored up in the stem than can be used up.

The pressure of the bark can also become harmful when the **natural shedding of the bark** of old trees is retarded. The development of the bark in the cases of those trees which present in their old age a rough bark is somewhat as follows: —First of all, the stem in its young and succulent condition is surrounded by a simple epidermis, consisting of parenchymatous cells. With the development of the woody cylinder a strengthening of this outer protective layer takes place, by the formation of several layers of tabular cork cells. When the twig has grown to a branch, and has therefore formed several annual rings of wood, these layers split the cork mantle open so as to be able to expand themselves. Close to these natural splits new cork layers are formed, which project like arches into the outermost layers of the cortex, and by cutting off such layers cause their ultimate death. These dead pieces are pushed outwards by the development of new bast and cortex, and project from the surface of the stem. The repetition of this mode of procedure causes the formation of the so-called **bark scales**. These scales, consisting as they do of dead tissues, absorb water during moist periods, and give it off again in dry weather, the cells contracting at the same time. By this contraction the dead scales become gradually separated from the living tissue beneath them. In this way, the stems protect themselves naturally against an excessive pressure of the bark, and the expansion or growth in thickness of the stem is able to take place. The scales drop off most readily when the weather is alternately wet and dry. If, in some cases, the tree retains its bark for too long a period, it becomes necessary to **scrape the stems**. The scraping of the stems should therefore be looked upon as one of the regular operations in arboriculture.

In those cases in which orchards have not been carefully treated (especially in damp situations), we find trees from which the bark scales have not been removed naturally or artificially for several years, and by decomposing have formed suitable places for an overgrowth of lichens or mosses. In such cases the pressure of the bark is greater than where the scales have been removed, and the growth in thickness of the trees is retarded. We may here begin by scraping the stems and then proceed to slit the bark along the whole length of the stem.

Sometimes we only desire to strengthen certain portions of stems or branches, where the growth in thickness has been retarded by the pressure of other branches or of stakes. What considerable tension is exerted upon the bark by the thickening of the woody cylinder can be gathered by observing the longitudinal slits some hours after they have been made. For while the wound inflicted was at the commencement only as broad as the back of the blade, a few hours after the operation the slits will be twice or three times as broad. The margins of the wound have become separated by the contraction of the strips of bark which were previously stretched. Such a small slit heals very rapidly, as the callus masses formed at the margins very soon meet over the cut. The cambium layers of the callus masses join, and soon complete the cambium ring which was destroyed by the incision. This cambium ring does not, however, at first form long wood fibres in the incisions and in their vicinity, but gives rise to a large amount of wood parenchyma, the cells of which are short and square. For this reason the growth in thickness of the stem is more rapid.

It is only when the wound has become covered in by a new cork layer, and when the latter has become continuous with the general cork mantle, that the pressure again becomes greater on the cambium cells, and the new wood elements assume an elongate form and become like the normal wood cells. Under certain conditions, the slitting of the bark, together with copious watering, may cause a renewed formation of spring wood, even after the autumn wood has begun to be formed. If the wound heals very rapidly and a long dry autumn sets in, then the ring marking the year's growth may show two layers of spring and two layers of autumn wood, and we have an appearance of two annual rings having arisen in one year.

The bark is sometimes slit in order to cause a thickening of a slender stem of a wild stock if the scion which has been inserted is very thick. In this case the numerous incisions which are made all round the stem commence in the scion close above the juncture of scion and stock, and extend some way down the latter; thus a considerable increase in thickness of the stock is obtained. In the case of grafts which have been made close to the ground, and in which the scion is strong and healthy but cannot develop properly, owing to the weakly condition of the wild stock, slitting the bark may be resorted to for the purpose of stimulating the scion itself to the formation of roots. In this case the numerous incisions are made in the bark of the scion alone at its lowest end, and the incisions are covered up tightly with stimulating soil. Roots should then soon make their appearance, and thus the independent nutrition of the scion would be commenced. This procedure would be applicable in the case of dwarf trees, such as pears grafted on quince stock, and apples on Doucin or Paradise stock.

The slitting of the bark should be carried out soon before the unfolding of the leaves, as the pressure of the bark is greatest at this period, and the releasing of this pressure will have the greatest effect upon the growth of the cambial zone.

If, however, the bark is slit in order to get rid of an excessive supply of water and nutritive material, then no special time can be recommended for the operation. It may be resorted to if at, or shortly after, the commencement of the vegetative period, a number of organs which would naturally have used up the nutritive material have been removed. A very well known instance is the disease of the gooseberry, which we have discussed in a preceding paragraph. The swellings on the stem of the wild stock below the point of insertion of the scion are formed because the water which the roots of the stock have been forcing into the stem during the winter is not sufficiently used up. For it is, unfortunately, often usual to cut away the lateral shoots too soon; consequently the bark is over-loaded with water, and the cells elongate in the region where they are youngest and where the pressure of sap is greatest, *i.e.*, at the uppermost portion of the wild stock. Thus a spongy swelling is formed, which bursts and then dries up, causing also the drying up of the graft. This unhealthy condition may be avoided by getting rid of the superfluous water and of the plastic material which was intended for the lateral shoots; both these results can be attained by slitting the bark.

Similar diseased conditions make their appearance, especially in cherries and peaches, if too many superfluous eyes are removed, or too many young shoots have their tips removed.

Swellings do not in this case make their appearance in the bark, but the cambium forms a very spongy wood tissue which is liable to gummosis. When this widespread disease is due to the above causes, it can be successfully treated by slitting the bark of the stem.

CHAPTER VIII

THE USE OF SHOOTS FOR PROPAGATING

§ 34. What is meant by layering, and of what use is it?

A NUMBER of shrubs have the faculty of readily producing **adventitious buds.** Such buds are not formed like the regular eyes in the young tissues at the tip of the stem, but take their origin at a later period, and usually in consequence of some external stimulus. The pith of these buds is not continuous with that of the shoot on which they are borne, but joins on to those layers which were the formative tissue or cambium in the year in which these buds took their origin.

Dormant buds, too, may grow out many years after the main shoot is formed, but they were formed in the same year as the shoot itself, and their pith is continuous with that of the shoot.

Such adventitious buds may arise on roots and grow out into shoots, and if separated with a portion of the root of the parent plant, become separate individuals (Raspberry, Lilac, some kinds of Plums, Cherry, Apple, Spirea, Rose, &c.).

On the other hand, many plants can produce adventitious roots from leaf and stem structures if they receive a certain stimulus. Gardeners use this faculty for purposes of propagation by stimulating branches or shoots to produce adventitious roots, and then separating them from the parent plant.

Sometimes by merely increasing the supply of sap a branch may be stimulated to produce adventitious roots. In other cases more potent irritation (such as is produced by a wound) must be set up to produce the desired effect, and it takes a longer time until the result is attained. In such plants, in which the formation of roots is difficult to bring about, the gardener endeavours not to damage the shoots which are to be experimented with, in case the operation should not be successful, and he tries, therefore, to produce a stimulus at various places on the shoot, by enveloping them with damp soil, but does not separate the shoot from the parent plant.

This process is termed **layering**. In certain cases this is done (in the case of tall and rigid plants, especially if they are valuable) by fixing some convenient receptacle to one of the branches at a certain height and filling it with earth and moss, which must be kept damp. After the branch has become rooted, it is separated below the roots from the parent plant. **But all methods of layering in which the branch remains in its natural position are less reliable than those in which the branch is artificially bent.**

This may be taken as a guide for all cases of layering, and the above-mentioned method of procedure is only to be resorted to when regular layering cannot be employed. We must remember, too, that the vigour of root formation in the same species or variety of plant depends upon the amount of assimilated food material stored up in that portion of the stem which is being stimulated by moisture. From what has been said in the chapter on pruning, we know that every bend, notch, kink, or ring cut on the shoot hinders the flow of food matter, and causes the plastic material to be stored up in the portion of the shoot above the wound. The bending of the branch therefore is in itself a favourable preparation for layering.

The actual process of layering differs for different plants, even when the shoots are all near the ground. If such shoots are not produced in sufficient numbers, they must be called forth by cutting off the mother-plant close above the soil, so that strong lateral shoots may be developed. These can be used already in their first year, when they are still herbaceous; they must then be bent outwards close to their insertion, and the stump of the mother-plant and all the bases of the shoots must be covered up with soil. If the branches have made but few roots in their first year, it is advantageous to ring the exposed places where they have been bent, and then to cover them up again. The tips of the branches which have now become woody are cut off, and the lateral shoots are nipped off, so as to let as much material as possible wander downwards, and not to let it be spent on the formation of new shoots.

The best time to cut down the parent plant is at the end of winter, as there will be no more frosts to damage the wound, and the dormant buds will soon begin to grow out.

This method is recommended for apples, quinces, and plums. In the case of the former, in many varieties there is sufficient root formation in the first year, while in the case of quinces, wood of two years' standing seems to be most advantageous. The branches may with advantage be twisted before covering them with soil at the point where it is desired for the roots to make their appearance. In the case of crab-apples, the quickest method of propagation is to cut in pieces the smaller roots, of about the thickness of a quill, which may have dropped off from the roots of a larger tree which has been transplanted, and cover them with soil. They very soon produce adventitious buds and form very fine young plants.

It is, however, more usual to bury the layered branches in the soil than to heap earth up round the plant. The parent plants, which are kept short, are surrounded by a trench, into which the strong shoots are bent down in such a way that their ends project far from the filled-up trench, and they are then cut back to two or three eyes. According to the slowness with which the branches form roots at the place where they are bent, they may be further helped on by ringing, twisting, or cutting. This method is employed in Gooseberries, Red Currants, Vines, Hazels, Quinces, and Mulberries, Elders, Guelder Rose, *Forsythia*, and *Magnolia*. Among Conifers it succeeds with *Juniperus, Taxus, Thuja*, and *Sequoia*. In France it is employed for *Aucuba, Rhododendron*, and other more delicate shrubs.

The long branches of climbers (*Glycine, Aristolochia, Lonicera, Clematis, Hedera, Tecoma*, &c.) may be arranged so as to form a series of arches, the intervening portions of the branches being covered up with soil. Thus between two covered-up pieces there is a leaf-bearing and food-providing portion.

The separation of the rooted portions should take place at the usual time of transplanting.

In the previously mentioned method of propagation by portions of roots with adventitious buds, as *e.g.* in *Rhus*, *Æsculus macrostachya*, and others, the production of buds on the roots can be fostered by continual pruning in of the mother-plant and by injuring the roots. But such wild stock produced from root buds is not so good for grafting upon as seedlings, as it has a great tendency to produce more adventitious buds, and therefore to weaken the main stem.

§ 35. What rules should be followed in striking cuttings.

A cutting is a portion of a plant detached from the parent stock, and which becomes an independent plant by the formation of new roots. The new roots take their origin either immediately at the cut end or at some little distance from the latter. In different plants the power of producing adventitious roots is very different. Speaking generally, we may say that the older the various organs of a plant are, the less inclined they are to form adventitious roots, and that of the various cultivated plants those are least able to be propagated by cuttings which have a hard and brittle wood. Of our fruit-trees, apples and pears very readily form a strong protective callus over a cut, but do not easily produce adventitious roots.

From the Paradise stock cuttings can readily be struck, however. The same may be said of the Myrobalan plum, Quince, Vine, and Ribes. Of other trees, soft-wooded Poplars, Willows, *Ailanthus*, *Platanus*, and *Paulownia* readily produce roots, while *Robinia*, Elms, Beeches, Oaks, Walnuts are not to do so.

If a cutting is to form roots, and therefore to develop new organs, it must contain a sufficient supply of plastic matter for that purpose. This material has either been formed in a previous vegetative period and is stored up in the shoot (**woody cuttings**), or the cutting must be able to form the necessary substances after it has been detached from the parent plant (**herbaceous cuttings**). The latter must therefore always be provided with leaves, while in the case of the former it is not necessary. Woody cuttings always form callus over their cut end; in herbaceous cuttings it need not be formed. The formation of roots in a cutting is not dependent upon callus formation.

We may here repeat again that callus is a thin-walled colourless tissue, consisting of meristematic cells arranged in close rows, of which the end ones are still in process of growth and which have not as yet become differentiated into cork or wood.

The first sign of life in a cutting manifests itself by an alteration of the tissues near the cut surface, the cut generally running obliquely across the shoot, and being close below a bud. If we cut off a shoot, we thereby expose all the tissues of which it is formed, and we then bring the latter in contact with a damp medium (water, sand, earth, sawdust, fibre, &c.). Some of the tissues which have been exposed are not able to form the healing layers of callus; this is always the case with the old wood, often with the pith and the outermost layers of the cortex. The layers which are capable of further division, and are therefore charged with the production of the protective callus, are the cambium, the very young wood cells, and the innermost layers of the cortex. The larger, therefore, the area of exposed wood as compared with the other tissues, the more difficult will be the healing process.

It is therefore essential for the success of propagation by means of cuttings to bring about a sufficient and natural closing up of the cut end of the shoot.

This closing takes place by two processes. In the older soft tissues (pith and old cortex) there will be formed above the wounded cells transverse layers of cork cells which protect the cutting against excessive moisture. The woody elements adjoining the damaged wood cells and vessels may become plugged up with a very resistant brown mass (gum) or with thylloses, which have the same effect in closing the apertures of these cells and vessels. The second process is the covering in of the cut by the formation of callus.

Both processes take place (with very few exceptions) more completely when the cut surface is richly supplied with air.

Care must therefore be taken that the medium in which the cutting is placed is **very thoroughly aërated.**

When the closing of the wound begins, the cells of the cambium, of the young layers of the wood and of the bast begin to absorb more water and to bulge out over the cut surface. When the delicate extended portion of the growing cell has attained a certain length, a transverse wall is formed behind the apex, and the latter grows on farther. As the cells, which bulge out in this manner and divide, are very closely set, long rows of cells will soon be formed which go on growing at one end, and being firmly packed together, form a delicate white tissue (Fig. 26). As long as this soft tissue continues to grow at one end, and thus increases its bulk, it is termed callus.

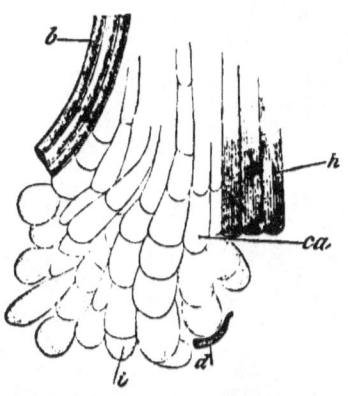

FIG. 26.—VERY YOUNG CALLUS, THE CELLS OF WHICH CONTINUE TO GROW AT THEIR APEX (*i*). PUSHING AWAY THE OLD AND NOW DEAD CELLS (*d*).
b. bast; *h.* wood cells; *ca.* cambium.

But, after a while, the plastic substances which are being passed down from the cutting to the callus find the path through all the callus cells to the extremity too long to traverse. The diffusion of the food matter does not take place sufficiently actively to the margin of the callus, and the growth ceases at that region. Instead of that within the callus, an arched strip of tissue makes its appearance, and its cells continue to increase in number. This strip of meristem becomes confluent with the cambium of the cutting, and represents the continuation of the cambium within the layer of callus. Within the latter it now commences to form bast cells on its outside, and on its inside new wood elements, and this forms the actual covering layer. The covering layers at the lower end of the cutting do not differ in any essential manner from those formed in the healing of a pruned branch, except that no green colouring matter is formed, owing to the absence of sunlight.

THE USE OF SHOOTS FOR PROPAGATING   175

Fig. 27 shows the process of healing in the case of a rose-cutting; s' is the line along which the shoot was cut; everything below this line is newly-formed tissue, or what is usually called callus. This callus has grown out in the form of a thick ring from the original cambium, and has spread from the margin over the cut end of the shoot. We can distinguish in this figure, which represents a longitudinal section, a ridge of

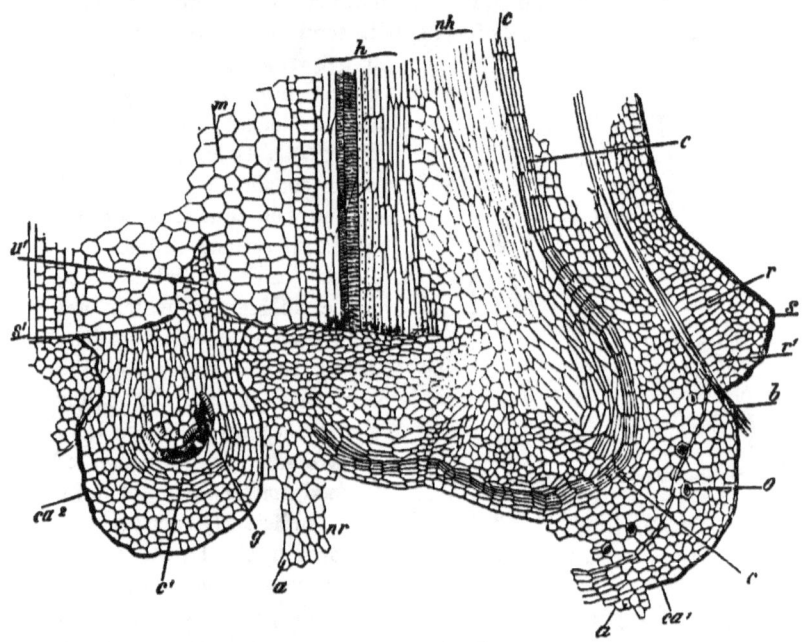

FIG. 27.—FORMATION OF CALLUS IN A ROSE-CUTTING.
(For explanation see the letterpress.)

callus (ca') cut radially, and a transversely cut ridge (ca^2) which runs towards the front and has already coalesced with the ridge (ca'). In this way the pith, it is true, has become covered in; but this has taken place without any active co-operation of its tissue, as it has simply been covered in by the fusion of the ring-like growths of callus, which have grown in from the circumference. In other cuttings, as, for instance, in the case of the Fuchsias, the pith takes an active part in the formation of the callus, and may, in some cases, form the greater portion of this tissue. In Fig. 27, furthermore, m represents the pith which has been cut by the knife. The groove u' is filled by the callus which is growing forwards from the back; h represents the old wood which was formed before the cutting was taken; nh, the new wood which has been formed since then. It commences with short, wide, porous, and thick-walled cells, filled with starch grains, among which there soon make their appearance some short reticulate vessels. These elements become more and more compressed towards the periphery, and resemble the normal wood more closely the nearer they are to the present ring of cambium, which means the later they have been formed after the cutting was taken. The cambium ring arches over the cut end, and is covered on the outside by the new bast and cortex, which layers are not shown in detail in the figure. At the outermost portion of the cortex will be observed the corky cells a, now disappearing, which were the first spherical or pear-shaped callus cells which covered in the cut. These rows of cells multiplied, as has been previously mentioned, by the constant divisions occurring in the terminal cells.

In that portion of the callus ring which is directed forwards (ca^2), and is therefore cut transversely, g represents the short reticulate vessels which are the commencement of the new wood. Surrounding the latter we find the cambium layer (c'); b is the old group of phloem formed before the cutting was taken. It has been pushed far away from the old wood by the formation of new cortical layers, and has died away at its free end. Those cells which adjoined the hard bast cells, however, have been liberated by the cut from the pressure of the bark, and have become radially elongated (r'), while in their normal condition their long axis was parallel to that of the shoot. The outer layers of the old bark (r) have not become changed: (o) is the oxalate of lime, which occurs partly in the form of single, partly in the form of clustered crystals. In some plants in which there is little tendency to produce adventitious roots, the formation of callus may become so dominant, that all the material which finds its way down the stem is used for the covering layers of callus, and the formation of roots does not take place (Conifers and Ericas). In these cases it is advisable to cut into the large knots of callus tissue. By this means the cambium layer receives a check, and a greater massing of the plastic substances takes place in the regions above the cut, and thus a stimulus is set up which may often result in the production of roots.

This method of procedure succeeds in cuttings (often leafless) taken from old wood, and has some similarity to the ringing of shoots. In the case of ringed shoots, the tissues forming the upper lip grow out into an expanding mass of cells, which cover in the exposed woody cylinder. If we imagine this woody cylinder to be cut through near the upper end of the ring after the production of this callus mass, and picture the latter covering in the cut surface, then we would have what actually takes place in a woody cutting.

In the case of shrubs with deciduous leaves, the best time for taking cuttings of woody portions is the beginning of the winter, or, in the case of shoots which will not be damaged by the frost, the latter portion of the winter immediately before the active growth of the spring takes place. Strong slips cut back to three or four eyes should be bound up in bundles and placed in a cellar, or only temporarily covered with soil, and when the spring approaches be planted in rows in a north aspect with only about two eyes projecting from the soil. In this way old wood which has been pruned away in the early spring may be used for cuttings. This at least answers in the case of *Rosa, Weigelia, Cornus, Deutzia, Lonicera, Ribes, Spirea*, &c.

The callus formation in the case of herbaceous cuttings is somewhat different. Generally more tissues take part in the healing process. It is here especially the pith which forms the bulk of the callus; the older cortical tissues may, however, be very active. Even the vessels of the wood may take part in this formation of callus (*Begonia, Thunbergia*), as the cavities of the vessels may become blocked with thylloses, which may grow out over the cut surface.

Propagation by means of such herbaceous cuttings is the most advantageous means of propagating plants, and this is true of woody plants too; only the treatment of the cuttings is quite a different one. We must always remember, that when the tip of a shoot is used as a cutting, its callus and adventitious roots are not formed from reserve food matter which has been stored in its tissues at some previous period, but that the materials necessary for these growths have to be formed by the cutting after separation from the parent plant. As soon as a portion of a plant is dependent for its food upon its leaves, we know that it needs light. **Herbaceous cuttings therefore need light,** and comparatively much light, while woody cuttings can do with very little light, at the outset at least.

The herbaceous cutting is taken before its axis is much lignified. The cut surface exposes tissues which have as yet no thickened cell-walls; the cells are rich in protoplasm and cell sap, are more prone to changes and decomposition, and require therefore an increased stimulus to continue the vegetative processes in spite of the wound which has been inflicted. This stimulus is provided by the increase of temperature. **Herbaceous cuttings require, therefore, more heat** than cuttings of the same species taken from older portions of the plant. In some cases indeed the temperature requisite for herbaceous cuttings is harmful for the woody cuttings, because it calls forth certain changes (possibly of a fermentative nature) the products of which cannot be used up at the time, and therefore cause decay.

We must remember, on the other hand, that the wound itself cannot be healed at once, and that the soft cuttings lose considerable amounts of water from their leaf surface by transpiration, and this at the time they are without roots, which could supply the requisite amount of water. We must, therefore, reduce in the first instance the transpiration without taking away the leaves. This can be done either by shading the cuttings or by keeping the air saturated with moisture. Every decrease of the amount of light diminishes also the amount of transpiration. In a damp or saturated atmosphere, too, the transpiration of the leaves is reduced.

Herbaceous cuttings require, therefore, at the outset a moist atmosphere. We purposely say at the outset, because it is a frequent source of error to continue this for too long a time. Absence of light and a large amount of atmospheric moisture reduces the assimilation of the leaves also to a minimum, and therefore decay often sets in on the cut surface and the cuttings are doomed. Herbaceous cuttings should, therefore, only be shaded during the first few days, and should very soon become accustomed to the normal illumination. After that allow the air to circulate among the leaves, and do not be afraid of the drooping of the leaves when the sun is shining on the cuttings. Gradually the cuttings will get accustomed to a more sunny and drier atmosphere, even if they have as yet no roots.

The great mistake in the treatment of herbaceous cuttings is to water them too much and to keep them too much closed in, in order to prevent them drooping. Even the most porous substance in which the cuttings are placed will act deleteriously if it becomes water-logged. The oxygen of the air is then prevented from reaching the delicate cut surface, fermentative changes commence in the cells, and the decay of the cut surface begins.

Herbaceous cuttings always require a well-aërated soil.

The art of the cultivator consists in executing these precepts in the way most suited to the individuality of the cutting.

No general rules can be given as to the amount of light, moisture, air, or heat necessary for cuttings, as the requirements of various plants differ so considerably. If, therefore, cuttings of many different plants are put into one propagating frame, some of the cuttings will very soon damp off. Let us just mention one or two of the most striking peculiarities. The scarlet Pelargoniums are so fond of light, and can dispense with water so well, that the cuttings may be put directly in the bed near the mother-plant. The same is the case with *Pelargonium grandiflorum*, which grows without much difficulty if the young tips are placed in a pot with sand in the sunny position occupied by the mother-plant. At the commencement they will droop, and the oldest leaves will actually dry up, but the young tips very soon recover, and remain turgid until the cuttings have rooted. If you tried a similar experiment with a cutting of the Heliotrope or Lobelia, the cuttings might easily be killed by a single hot day. They will, however, grow in a warm saturated atmosphere which would cause the decay of the tips of the Pelargonium.

Cuttings of the double white Primula can do with a good deal of moisture and of warmth, but they must have a large amount of light. Those who have no bottom heat at their disposal should keep the cuttings fairly dry, and can place them in the open in a fairly shady position. The leaves, it is true, will wither, and the rooting will take a long time, so that a gardener could not use his method with any degree of security.

No time can be fixed for striking herbaceous cuttings. Shoots may be taken as soon as they have received a certain rigidity and have some fully developed leaves, supposing always that there is still sufficient time favourable to vegetative growth to enable the formation of roots to take place.

If, however, the season is too far advanced to make success certain, it is better to take fully developed leafless woody shoots for propagating. Besides this kind of woody cutting mentioned, which may be called a winter cutting, in some shrubs not liable to damage by frost autumn cuttings may prove successful. Less hardy genera, too, if they are well protected from the frost, give very good results with autumn cuttings (Roses).

We may give the following general instructions for the treatment of all cuttings made from dormant wood.

We want the shoot which we have put into the soil to use its starch and the other stored up food material for the production of callus and adventitious roots, instead of passing it into the buds, as would have been the case if the branch had been left attached to the plant. The conditions in which the cutting is placed are very favourable for such a course; for on the one hand, the wound caused by the cut, when in contact with the damp soil, acts attractively to the food matter, which has become soluble during the vegetative period. Secondly, one main factor in the expansion of the buds is wanting, and that is the root pressure. The buds of cuttings, therefore, unfold much later than those of the parent plant.

These conditions must, therefore, not be disturbed, but should be maintained until the callus covering the cut is able to absorb sufficient water to give to the buds the necessary turgidity, and to push forth the stem apex and the young leaves. To commence with, therefore, we should ensure a complete rest of the upper portions of the plant. These conditions obtain especially in the autumn, when the air is greatly cooled and the soil is still warm. If cuttings are taken in the spring, it is advisable to plant them in a northerly aspect, so as to prevent them from becoming heated, and to prevent any premature development of the buds. In the case of cuttings set in cold frames, the glass or other covering should be removed as early as possible in the spring.

The neglect of these precautions is the cause of most injury to cuttings kept in cellars, greenhouses, &c. The warmth of these places stimulates the buds to premature expansion; the new shoot will then form the chief centre of attraction for the reserve food substance, and will grow out as long as the food supply lasts. Thus the formation of roots is prevented and the cutting dies from want of root nutrition.

In the case of valuable plants, which do not readily produce adventitious roots, preparatory steps may be taken by ringing or constricting the branch which is to serve as cutting, and not separating it from the parent plant until the callus ring is formed. The tip of the branch is then removed so that the cutting has only three or four eyes.

In certain cases the stimulating effect of the warm spring weather acts beneficially. Quite a number of shrubs (*Weigelia, Deutzia gracilis, Cydonia japonica, Euonymus japonicus, Aucuba, Buxus, Laurus, Andromeda, &c.*) grow best from cuttings made from the completely ripened first year's shoots separated in July or September from the parent plant.

The cuttings may be placed in the open ground in rather shady positions in sandy soil, or may be placed in boxes which remain in the open. During the first few weeks it is well to keep them out of direct sunlight; later on, towards the autumn, they may receive plenty of light. In this method of propagation most of the cuttings will have formed roots by the time the winter sets in; those which were taken very late in the year will only have the protective callus. Of the latter cuttings those which are in pots or boxes should receive a little bottom heat in the spring; the roots will then soon make their appearance. This is especially advisable in the case of *Thujopsis, Thuja, Cupressus, Juniperus, Sequoia*. The more delicate of these should be slightly forced, so as to ripen their wood sufficiently by the end of June; still the cuttings will in most cases only have formed callus by the autumn, and **will only produce adventitious roots in the spring when stimulated by bottom heat.**

The peculiarity of the cutting to continue the special mode of growth attained by the parent plant can and is made use of to form new varieties.

Variations, such as variegated plants, fasciated specimens, or double-flowering varieties, can thus be perpetuated by cuttings. Of special interest are the curious juvenile stages through which some Conifers pass, and which have been fixed by cuttings and brought into the market as new species and genera. Thus it is mentioned that to obtain *Chamæcyparis squarrosa*, cuttings must be made of the small shoots of *Biota orientalis*, which appear immediately above the cotyledons, and which bear cruciate leaves. In the same way cuttings of the first lateral branches of *Callitris quadrivalvis* give rise to a quite new form. The fixed juvenile condition of *Cupressus sempervirens* may possibly have produced *C. Bregeoni*; the first shoots of *Cupressus Lawsoni* have not got imbricate, but horizontally expanded leaves. *Retinospora ericoides* was obtained from *Chamæcyparis sphæroidea var. Andalyensis*, &c.

Root-cuttings have been dealt with in the paragraph on layering. They are portions of the underground axis, which are stimulated by being cut in pieces to produce adventitious buds. The shoot which arises from such a bud becomes an independent plant of strong growth as soon as it produces adventitious roots from its own tissues. *Ailanthus, Aralia, Paulownia, Rosa, Pirus Malus, Mespilus*, and others can be propagated in this way, if strong pieces of their roots about two inches in length are cut off either before the first shoots are formed in the spring, or before the second shoots are produced (July). These pieces are then planted in rows in the soil. Among Conifers, *Araucaria*, *Podocarpus*, and *Gingko* are mentioned as suited to such methods of propagation, especially if some bottom heat can be used.

In the case of some plants (*Vitis vinifera* and *Pæonia arborea*) a special and advantageous method of propagation is practised by means of **eye-cuttings**. The buds are cut out of the old wood in the spring, in such a way that they resemble eyes used for grafting, attached to a small piece of wood, and the latter is placed with its cut surface downwards in pots in a sandy soil. The pots are then sunk in a hot-bed.

In plants, too, in which the axis is succulent, the eyes can be cut out for propagating purposes (potatoes, &c.), or the whole tuber may be cut in pieces. In this case we might speak of **tuber-cuttings**. Their further growth is, as a rule, a certainty, especially if the precaution is taken of keeping the cuttings for a few days in a dry place before they are planted in the ground. By so doing the formation of a layer of cork cells immediately below the cut surface is brought about, and this constitutes the best protection of the soft and succulent tissues against decay.

§ 36. What object have we in view in budding and grafting, and how are these operations best performed

In budding and grafting, either one or several buds of the mother-plant are detached and inserted on an older stock of the same or of a similar kind. By this insertion of a younger portion of a plant on an older stock the former can reap all the advantages of the more advanced age of the latter, becomes indeed older itself.

Both processes necessitate the cutting of the tissues; that method will, therefore, be the best one in which the healing of the wound takes place most rapidly and most completely.

The rapidity of healing depends (other things being equal) upon the relative extent of the cambium, the layer which will produce the healing tissue, as compared with the entire cut surface. The greater the amount of cambium exposed on the cut surface, the better are the chances of the graft or bud.

From this point of view **budding** is the most advantageous of the two operations, as it consists in the insertion of a bud directly upon the cambium ring of the wild stock. To reach this cambial region the bark is slit open by a T-shaped cut, and the bud is placed under the raised lappets of the bark. The bud may be attached either to a shield-shaped mass of bark alone, or there may be some wood adhering on the inner surface of the latter. In that case the operation is spoken of as "**budding with wood.**"

In the case of budding, the healing process can practically take place at all the points of contact. As the bark of the old stock has been split in the cambial region, the youngest splint-wood remains on the surface of the wood, the youngest bast cells line the lappets of the bark. From both the regions normally new layers of cells will arise, which tend to fill up the interstices between the scion and the stock. Later on the scion itself will take part in the healing process, sending out similar callus like rows of cells from its inner surface, just as was done from the bast of the wild stock.

If the bud has some wood attached to it, the healing process can only take place by means of the narrow cambium zone bounding the scion. In this respect budding with wood is less favourable, as the scion offers less surface for fusion of tissues to take place. But this is of little importance in the case of strong wild stock, as it will form healing callus so rapidly that the participation of the scion in this process may be neglected. On the other hand, this method of budding is much more easy in the case of wood from which the bark cannot readily be peeled, and is therefore much more successful at the hands of an unskilled operator. For it often happens in separating the bark from the wood that the fibro-vascular bundles of the bud remain attached to the wood in the form of a small conical protuberance, and the bud is only represented by a hollow cap, which does not grow out even when the bark has united to the stock.

How rapidly the fusion of tissues takes place in the case of budding can be seen by the changes which soon make themselves apparent. Already after twelve hours the cut margins of the bark, and the outermost wood cells, show a thickening of their cell-walls, the contents of the outermost layers of cells increasing in amount.

But it often happens that all the cut surfaces do not take part in the healing processes. If the splint-wood does not show any new cell division, its cell-walls swell up and turn brown; the outermost cells of the splint-wood then collapse and form a thick irregular brown band. This discolouration is, however, quite slight or does not take place at all if the cells remain capable of division, and its formative power is so great that already after two days (in the case of the Ash) several layers of callus cells will be formed.

In the case of the lappets of bark which have been raised from the wood the cells actually forming the cut margins usually die away; those beneath them may protrude a little, but rarely do more than form protective layers of cork, while the tissues closer to the base of the lappet show active cell division and the formation of healing tissue. But though the exposed layers of cells within the wound only form a moderate amount of new tissue, this is as a rule sufficient to cause a complete temporary closing of the wound. We say a temporary closure, because, as a matter of fact, this first formed tissue only lasts for a short time. For as soon as this healing tissue has attained a certain development and is exposed to a certain pressure, a zone of meristematic tissue arises within it, which unites with the cambium layers which grow out from the angles of the lappets of the bark. This meristem now begins to form the woody parenchyma which covers in the first formed thin-walled closing layers, and which gradually becomes filled with starchy reserve material.

Finally, the cambium layer of the scion unites with the continuation of the cambium of the stock, and thus a new continuous cambium ring is formed, of which the scion forms an integral part. The transplanted bud has united with the stock and behaves as if it were a bud of the latter (Fig. 28).

In the adjoining figure, representing a transverse section through a stem of a Rose upon which a bud has been inserted, these conditions are all represented. W is the wild stock; RL are the lappets of bark, which have been raised by the T-shaped cut; E is the bud which has been placed within these lappets.

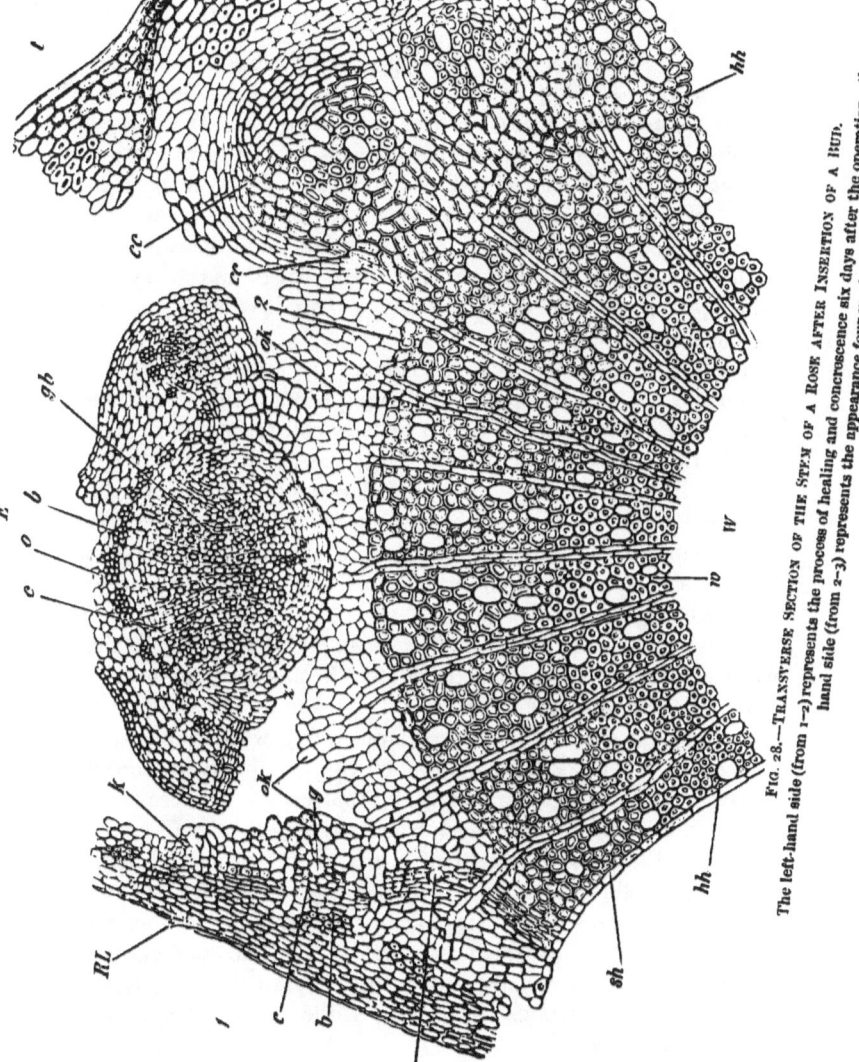

FIG. 28.—TRANSVERSE SECTION OF THE STEM OF A ROSE AFTER INSERTION OF A BUD. The left-hand side (from 1-2) represents the process of healing and concrescence six days after the operation; the right-hand side (from 2-3) represents the appearance four weeks later.

In the case of the stock, hh represents the old wood of the previous year, sh that of the current year, which was formed before the budding operation was performed. In the bark, b represents the bast fibres and t dead cells of the cut margin. In making the T-shaped cut, the separation has occurred in the youngest layers of the wood, a not infrequent occurrence in practice, and the youngest wood vessels (g) and the cambium (c) have remained attached to the bark. But often the reverse takes place, and the cambium and young bast cells remain attached to the wood.

Already, after twelve hours, a change has taken place at the margin of the cut, the cell-walls having become thickened and assumed a yellow colour; very soon new cell formation begins, and in the case of strong stock with the formation of callus (ok) in the splint-wood. On the lappets of the bark the marginal cells full of protoplasm (k) first bulge out, and their activity increases the nearer they are to the angle, where the bast is still attached to the wood, until they form a considerable amount of callus (ok).

These masses of callus thus formed by the bast and the wood and (under favourable conditions) by the scion unite into a temporary covering for the wound. Later on the activity of the cambial layer of the cortical lappets (cc), which represents the continuation of the cambium of the uninjured portion of the stem, becomes more pronounced; this leads to the formation of the actual and permanent wood parenchyma (kg) which represents the commencement of a growth of wood which continues under slight amount of pressure. This parenchymatous growth of wood, as it increases in amount, gradually crushes up the thin-walled tissue which was first formed for the closure of the wound. Gradually the whole of the tissue (ok) lying between 1 and 2 is replaced by cells of the nature of kg, which are filled with starch.

The scion (E) is in this figure a bud with an attached portion of cortical tissue, but without any wood cells; the actual bud lies in the direction of o, somewhat higher than the section represented in the figure, which only contains the large fibro-vascular bundle (gb) belonging to it. On the right of the chief bundle we see another smaller one, the corresponding one to which on the other side has been removed. The two smaller bundles are of no importance for the success of the operation, but the large median one is of great consequence. It is the small protuberance which must be present on the inside of the bud, as every operator knows, if the bud is to develop. If its place is taken by a small depression, then the vascular bundle (gb) belonging to the bud, and which has to form the woody cylinder of the new shoot, has remained attached to the parent plant. The bud then is represented by a hollow cap, which will not develop any further, even if the wound should heal up.

The union of tissues takes place most favourably if a layer of callus is formed over the entire inner surface of the scion, as is shown in the figure. The cambium zone lying beneath the bast fibres (b) has formed new cells most profusely, as is shown by the projecting group of cells at z. The cambium layer of the scion (E, c) unites later on with that of the stock (cc), and thus a continuous cambium cylinder once more exists round the stem, and the scion appears inserted in the tissues concerned with the normal nutrition of the stem.

The method of union of the tissues which takes place after a successful operation we need not describe minutely, for we know how an obliquely cut branch becomes closed in by callus. The cambium layer produces at first soft callus-like layers, which afterwards become hard and cover in the exposed woody tissues. The scion behaves like a cutting, taking the material necessary for its callus formation from the reserve food material which is stored up in its tissues. The stock, which is cut off obliquely above the point at which the budding has been effected, draws the material necessary for closing this cut from the surrounding cells, the cut having been made near to a bud which will soon grow out. The shoot produced by this bud will soon form new plastic matter through the assimilating energy of its own leaves, and the cut will soon be completely healed over.

But the cut surfaces of scion and stock are in close juxtaposition; the callus tissues formed by them must therefore very soon meet and press one upon the other. The effect of this pressure is to cause the fusion of the callus cells of scion and stock, the cambium layers uniting to form a continuous cambium cylinder. The closer the cut surfaces are one to the other, the more rapidly do the cells which are capable of fusion meet, and the more quickly and completely is the union brought about. Hence gardeners have adopted the rule that in all **processes of grafting and budding the cambium layers must be brought into contact.** This, however, is only then completely possible when the scion and stock are of the same thickness. If this is not the case, we must at all events endeavour to secure that on one side the two cambial layers will meet, so that the scion may from the outset be nourished, and that it may at least receive the water which it needs.

If the scion becomes attached at one side, at all events, so that it does not suffer from lack of water, it will soon be able to help itself. The growing buds will form such quantities of food matter that it will very soon form large masses of callus, and these will cause a union to take place on those sides which do not fit. This is most frequently seen in the case of cherry-trees which have not been properly ligatured. The scion will in these cases often be found to have an oblique position owing to the fact that on the non-fitting side large masses of callus have been formed which did not at first unite.

In such cases the advantage of properly binding up the graft is seen; the callus masses must not be able to push each other away. If the scion cannot give way, the callus layers will, under the action of the pressure, unite. The separating of scion and stock after the operation can be prevented by the so-called **tongue grafting** or tonguing.

The fusion of stock and scion takes place with the greatest difficulty in the case of **crown grafting.**

The dangerous element here is the wound which is inflicted upon the wild stock, which is either split longitudinally or has at least a wedge-shaped portion of wood removed from its upper end. Now the two cut faces of the wound expose the old wood, which cannot give rise to any healing tissue. The closure of the deep split, therefore, can only be effected by the ingrowing of callus from the peripheral cambium layers. This, however, only very rarely takes place completely; generally there remain gaps at the centre of the stem, and the tissues which immediately surround these become brown and die away. If water finds entrance to these cavities, decay very soon sets in with far-reaching effect. Different kinds of trees behave very differently when crown-grafted, and in certain genera this operation may be undertaken without the slightest danger, because they have (as, for instance, in the Lime) the faculty of forming callus cells by the renewed activity of pith cells. In such cases the callus will be found making its way outwards through the splits, and fusing with the callus formed by the cambium.

Such an instance of complete union of stock and scion in the case of a crown graft is shown in Fig. 29, which represents a transverse section through a two-year-old stock of the Lime, *Tilia Europæa*, upon which a scion of the Silver Lime has been grafted.

The operator had split the stock (*k*) very completely, had cut the scion (*Pf*) in such a way that some bark remained attached at both sides, and had forced it right into the region of the pith of the wild stock. The brown margins (*s*) denote the boundaries between the stock and the scion. Throughout the section *hh* stands for the old wood formed in the previous year, *nh* for the new wood formed during the year in which the grafting was performed, *g* for the wood vessels, *c* the cambium layer, and *mz* for the normal thick-walled, unchanged cells of the pith: *rp* is the normal parenchyma of the cortex and phlœm, with *b* the concentrically arranged layers of hard bast; *rc* is the commencement of the callus formed by the cortex.

As soon as the scion was placed in the split, the tendency to close up the wound made itself manifest. The damaged cells of the cortex died off, and became separated from the still living ones by a layer of cork cells. The young cortex, the cambium, the young wood, and here also the outer cells of the pith, commenced to divide rapidly and formed callus tissue, which forced itself in between the stock and the scion. The arrows in the tissue indicate the direction in which the callus masses moved, and the dark lines denote the line of juncture of the young healing tissues. The chief amount of this tissue is made up of the callus *rc*, formed by the cortical tissue; but both

THE USE OF SHOOTS FOR PROPAGATING 191

the young wood (*nh*) and also the pith have taken part in the healing of the wound. The latter has, for instance, formed

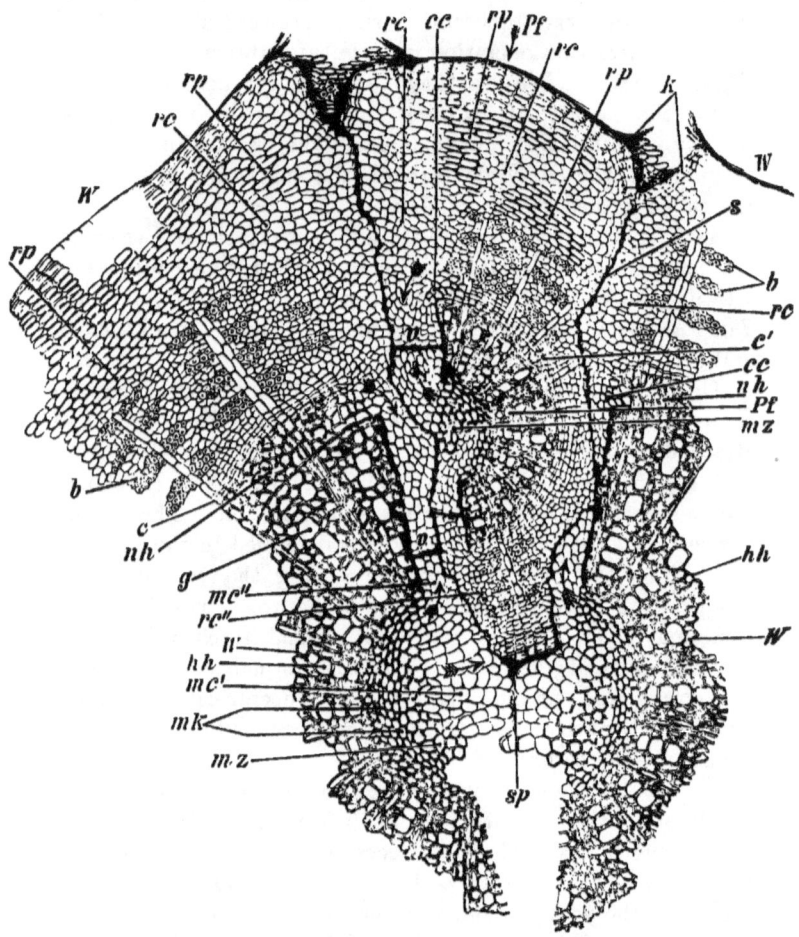

FIG. 29.—SECTION OF A GRAFT OF A TWO-YEAR-OLD TWIG OF THE LIME (*Tilia Europæa*). The scion (*Pf*) has been pushed into the pith of the wild stock. The arrows indicate the direction in which callus layers have grown to close up the wound.
(The explanation of the lettering will be found in the letterpress).

the tissue (*mc″*), which has joined up with that formed by the cortex (*rc″*). That part of the split (*sp*) which was not filled by scion has become closed by the activity of the pith (mk), which has formed a number of rows of callus cells (mc'), which continued to grow at their apex, until they met those produced from the other side of the pith.

Thus we see all the cavities between the stock and the scion have become completely filled by callus, and as the cambium of both c and c' have fused, we see that crown grafting may be performed, in the case of the Lime, as readily as any other method of improving the stock.

The healing is also very successful if the scion is not cut into a triangular wedge, but if it is only pared down on either side so that both sides are still covered with cortical tissue. The centre of the split is in this case filled by cortical cells which can form callus tissue to fill up the crack. Crown grafting cannot, however, be entirely neglected. This operation has to be resorted to in the case of old stems, the cortex of which does not easily separate from the woody cylinder, and it is advantageous when grafting on root-stocks (*Pæonia arborea, Clematis, Bignonia, &c.*). After the discussion of the three chief methods employed for the improvement of wild stock, we may now discuss the advantages of certain modifications of these methods.

Thus small graft shoots may be inserted into a T-shaped incision into the cortex of the stem. This is most advantageous in the late summer. The shoot is then trimmed like an ordinary graft. This method is very advantageous, and generally practised in the spring, as soon as the cortex can be easily separated in the stock. It is of advantage in the case of trained trees in which some branch has been injured, and also in the case of an old stock which is too thick to allow of a bud being introduced into it. This method might be called budding with a shoot.

If the crown of the old stock is to be preserved, lateral grafting may be practised. In the case of evergreen shrubs (Camelias, Rhododendrons, Azaleas, Conifers, &c.), a last year's shoot still provided with its leaves is used as a scion, and a piece of the stock of similar size is cut away from the side of the stem, so that the wound which is thus caused is covered as completely as possible by the scion.

Instead of removing the piece of cortical tissue of the stock, however, it may be left connected with the stem below, so that the wound is really only an oblique slit in the stem. In this slit the wedge-shaped end of the scion is fitted and bound up. It is still better to make the slit in such a way as to have only one cut margin. This is done by pressing the knife down into the stem below the cortex, in such a way that its point does not project through the cortex on the other side. Into this tangential slit the pointed end of the scion is then inserted.

In the case of thicker stocks and slender scions, it is best to substitute crown grafting by some form of graft in which the scion is inserted into the cambium zone of the stock plant.

How should grafts be bandaged? This question is often put, and is answered in various ways. To form a true judgment on this point, we must consider what is the object of binding up the wound, and also what material it is we employ. We have already stated that energetically growing grafts do not in some cases unite, because the callus masses formed by stock and scion respectively tend to keep them apart. The ligature has the function of preventing this, and of pressing the scion against the cambium of the stock. When the fusion has taken place, and the plastic matter formed by the leaves of the scion begins to take its normal downward course, then the thickening of the whole axis begins. If the material employed in binding up the branch is sufficiently elastic, then no stoppage or retarding of the thickening will take place at the graft. But if we make use of very closely spun wool or of bass, the ligature will soon be seen to constrict the cortical tissues, and this should be prevented. From this it follows that it is better to bind up grafts with flat than with twisted bands, and that cut surfaces which are concave to one another should not be laced up to meet. Such concave surfaces, however, should not and do not generally occur in the case of skilful operators. Novices will do well to prevent the transpiration from such badly fitting grafts by the use of a little wax. If the stock and scion are full of sap, if the operation has been skilfully carried out, and the ligature is firm but not constricting, the time involved by coating the wound with wax may often be saved.

§ 37. To what extent do scion and stock mutually influence one another?

As long as it was supposed that cambium was a sort of sap which circulated between the bark and the wood, it was firmly believed that the characters of scion and stock became gradually intermingled. Now, however, we know that the cambium is a tissue, the young cells of which have inherited from their first formation the tendencies of their mother-cells, and therefore continue to function in the same way, forming the same sort of cell-wall and cell-contents as their predecessors did. The characters of the scion as well as those of the stock will develop themselves separately in their several tissues.

But it cannot be denied that in those places where the callus cells of scion have united with those of the stock, the cells adjoining each other and belonging to two different plants will influence each other to a certain extent. Each cell prepares its substances in its own peculiar way, and those of its contents which diffuse out into the neighbouring cells, or, in case of the sieve tubes, pass over in a mass, will cause a difference of nutrition, and such a difference of nutrition will be connected with differences of character. It is, therefore, not strange that we should be able to demonstrate the influence of scion and stock one upon the other. But a fusion of characters never takes place, because the main functions of each cell are too well fixed by heredity to be able to be changed by mere differences of nutrition.

But apart from these considerations, the grafting will entail a certain mechanical effect, because in the new individual, consisting of the two parts, a transverse layer of short wood parenchyma has arisen.

This layer causes a certain amount of difficulty to all conduction. The wood vessels of the wild stock, as far at least as they were formed before the grafting was effected, are not directly continued into the scion, because this layer of callus lies between them. And even when the fusion of tissues has taken place, and the continuous zone of cambium has formed uninterrupted vessels which pass through the juncture of the two individuals, they retain for many years a sinuous course which retards the passage of substances.

The changes which take place by grafting cause for a number of years a retardation of the transport of food material.

But several examples may be mentioned of apparent changes brought about by grafting, one of the most remarkable of which is the transmission of a **white-leaved condition** (*albicatio*) from the scion upon the stock, and also *vice versâ* (*Passiflora, Jasminum, Abutilon*).[1]

It is less common to find that the occurrence of red cell sap in the leaves of the scion has extended to the wild stock. This phenomenon has, however, been noticed in the case of the copper-beech and the red-leaved hazel. The statements as to any effect produced on the flower of the wild stock are very few. A white flowering Jessamine upon which a yellow flowering variety was grafted is said to have occasionally produced yellow flowers.

But it is very well known that certain stocks have a very pronounced effect upon the habit of growth of the scion. Apples grafted on Paradise stock or dwarf stock (*Pirus praecox*) remain of short stature and often produce flowers in the year following the grafting. Grafted on the Doucin the varieties

---

[1] This occurrence may perhaps, however, be attributed to mechanical influence. The appearance of white patches or of entire white leaves depends upon the fact that in several portions of the tissues the cell contents appear poor; no chlorophyll grains are developed, and the protoplasm appears as a colourless flocculent mass. The cells have, therefore, received too little plastic material during their formation, and have not been able to increase the amount they had received. But it is possible in some plants which have a tendency to white foliage to induce the formation of certain white-leaved shoots by a continued pinching back of the young shoots. This may be explained by the assumption that these shoots are induced to develop before they have sufficient plastic matter, and had to develop very rapidly. Now, grafting acts like pinching, and stimulates the lateral buds to growth before they have sufficient food material. The leaves expand, and in the strong illumination soon become old, and cannot produce new food material for the impoverished cells, which therefore remain without chlorophyll. This character would not be transmitted, but would be brought about by the removal of the apex of the shoot in the operation of grafting. This also would serve to explain why the appearance of white-leaved shoots does not always take place close to the graft, but often at some distance from the latter, while green-leaved shoots are found close to the point of union.

become bigger and fruit later; on the crab-apple the tree retains a normal growth, but the crown does not produce flowers for a considerable time. In the case of pears, the quince and the damp-loving hawthorn will form dwarf stock. Trees with such stock, however, are known to have a shorter duration of life than when true wild stock is used.

Quite a number of further statements exist, which we shall partly repeat here, not so much to put forward these cases as of general moment, but rather to give an impulse to new experiments in this direction. Unfortunately there is a great want so far of systematically arranged experiments.

It is generally held that tree-like forms succeed better on bushes than the reverse. Sour cherries grafted upon sweet ones succeed less well than the latter upon the former. Cabanis states that late varieties of walnuts and chestnuts never succeed upon early varieties; while in the case of pears and apples, &c., this method of grafting late varieties upon early ones is said to have very good effect. The same holds good with peaches, and causes an earlier ripening of the fruits. Almonds on plums and the reverse will readily grow, but seem to deteriorate after a few years. The almonds grow more rapidly and commence earlier in the spring, and they also form a very large callus growth. It may be safely assumed that a very early scion, which requires constantly more water than the stock, will succeed upon a less vigorous stock only so long as the stock is able to supply the requisite water. If the scion cannot suit itself to the stock, it will soon perish from want of food matter. But the nature of the soil, the water supply, and the variety are three factors which cause a great amount of variation in the results. On the other hand, a stock which begins to function early in spring and forms a large amount of wood will supply a more or less exacting scion with more sap than it will be able to make use of. The superfluous material of the stock will tend to the rapid production of new growths. If the wild stock possesses many buds, its energy will expend itself in the development of long shoots; but if, as is generally the case, the lateral branches and eyes have been removed, the superfluous material will become lodged in the cambium of the stem, and will cause the formation of large masses of wood parenchyma in the place of the normal wood cells. This parenchyma, however, is more delicate and sensitive to external conditions, and in the case of stone-fruits liable to cause the exudation of gum. Indeed, Duhamel has already pointed out that branches of the almond upon which shoots of the plum had been grafted were liable to gummosis at the juncture of stock with scion.

If pears are grafted on quince or apple on Paradise stock, the death of even healthy shoots of the scion will take place if the soil is at all dry or there is an insufficient supply of roots. In one case it has been proved that the injury to the roots caused by transplanting was sufficient to cause the death of the graft. Grafts made at the same time on similar trees, which were, however, not transplanted, turned out quite successful. Peaches grafted upon plums do not form successful grafts. The scion forms a reddish wood and soon dies off.

It is, however, remarkable that pears and apples which result in successful grafts with rather distant forms do not unite well if the stock and scion are nearly related. Apples unite very well with pears, and the apple scion bears fruit at a very early stage, but it dies very soon. Pears grafted on apples also yield fruit for a time, but the scion very often shows an excessive formation of wood. (This depends, of course, largely upon the tendencies of different varieties to form wood.) This often results in knotty excrescences near the graft, in the premature dropping of fruit, or the production of flowers without fruits, and in a crippling or complete dying away of the crown.

Evergreen scions are quite able to unite with stocks which lose their leaves annually. Slips of the Cherry-Laurel (*Prunus Laurocerasus*) are said to unite with the stock of the Bird-Cherry (*Prunus Padus*). *Quercus Ilex* has been successfully grafted on *Quercus sessiliflora*, and *Cedrus Lebani* on *Larix europaea*. We have, however, no observations of the grafting of species which lose their leaves on evergreen stock.

Lastly, we must mention the fact that hybridisation has been occasionally brought about by grafting. Many instances of **graft hybrids** will be found in botanical literature, but many of the instances are of doubtful authenticity. But the possibility of modifying certain characters which are easily liable to changes (colour of flowers, &c.) by grafting must be considered a reasonable one, and we must admit that there are some forms intermediate between the stock and scion which may have resulted by a form of hybridisation.

CHAPTER IX

THE TREATMENT OF LEAVES

§ 38. What is the effect of injuries to the leafy tissues?

From what has been previously stated with regard to the work done by the leaves, it follows that every injury to the leaves means a decrease in the amount of new material formed. The removal of the smallest amount of leafy tissue entails the loss of some of the cells which should build up new organic matter.

As the veins of each leaf lead downwards through the leaf-stalk into the shoot, and as the soft bast and bundle sheath conduct downwards the organic matter formed in the leaves, the diminution in the organic matter formed will first make itself felt in that portion of the shoot which is close to the insertion of the leaf. If the shoot still requires the food material for itself, the injury to the leaves will cause the shoot to be less completely developed, *i.e.*, will cause it to remain short.

This is the explanation of the success which accompanies the removal of the tips of leaves of a shoot before it has done growing. The removal of the upper portion of the leaf causes the adjoining internode to remain short. Peaches regularly produce premature branches, *i.e.*, the buds which have just been produced in the axils of this year's leaves grow out immediately. These premature (*proleptic*) shoots have the peculiarity that their lowest internode grows several inches in length instead of remaining short, as it does in branches which are produced after a long resting period. The lower leaves and eyes of these premature shoots are, therefore, also several inches from the parent shoot. That is, however, a fault in trained trees in which one of the objects to be attained is to have the fruit as close as possible to the leading shoots. Now, if the tips of the lowest leaves are removed, the internodes belonging to them remain short, and the eyes will be in proximity to the leading branches.

Recently this method of procedure has been adopted in the case of herbaceous plants (Dahlias), and short, bushy plants are said to be produced.

The fruit-grower is sometimes forced to a partial removal of leaves to ensure a better development of his fruits. In summers with little sunlight, the almost entire removal of the leaves will be beneficial to fruits which have already attained the requisite size but need the ripening effect of the sun.

A considerable amount of defoliation is often useful in the case of trees growing in a rich damp soil, to accelerate the ripening of the wood. In such soils after a dry summer a wet autumn will often prevent the formation of winter buds, and the shoots will go on growing until December. Such shoots, however, are very delicate, as they have little wood formed, and are liable to be nipped even by moderate frosts. By pinching off the apex of the shoot and removing some of the leaves, the ripening of the wood is considerably accelerated. The removal of the apex of the shoot removes the tissues which attract the greatest amount of water, and the removal of the leaves allows the shoot to receive more light, which is one of the chief factors in ripening the wood.

We must now also consider those cases in which the foliage has been damaged against our wish by climatic or other disturbing influences. In such cases the amount of damage done depends largely on the cause of the injury. The most detrimental are the damages done by cockchafers or caterpillars. This generally takes place when the leaves are still young. The tree has expended the greater part of its reserve substances on the production of foliage, and has as yet received little or nothing from them. Once more it is leafless, and it requires more reserve material for the unfolding of those buds which have remained uninjured. It has now to draw upon the material stored up more deeply in the main stem, for during the first unfolding the shoots and the roots have used up all the material stored up in their vicinity. After the first production of foliage there remains still a considerable amount of starch stored up, chiefly in the base of the stem. The formation of a second crop of leaves will use up the greater part of this reserve material. By so doing the stem will be greatly exhausted. There will also be no material at hand, for a time at least, for the growth of the cambium, and consequently only a very weak annual ring of wood will be developed.

It is probably less harmful to the tree to lose its foliage by a late frost. The leaves killed by the frost remain attached to the tree, and it is therefore probable that a certain amount of the mineral substance, at least, which has been expended on their production will be reabsorbed by the branches.

Such a reabsorption of substance from the leaves by the stem has been proved in the case of foliage killed by excessive heat and drought. This takes place when young trees are growing in a very thin layer of soil covering rocks or scries. The dried-up leaves will remain adhering to the branches throughout the winter.

The nitrogenous constituents and the phosphoric acid, it is true, remain in the dead leaves, but the starch, and with it the potassium, is in part reabsorbed by the tree.

§ 39. In what cases can the leaf be used for propagation?

As has been mentioned in dealing with the shoot, some plants have a considerable tendency to produce adventitious roots and adventitious buds. This faculty is in many cases shared by the leaves too, and causes them to be used as cuttings. At present we know of no general rule as to what kinds of leaves can be used for cuttings, and we can only base our remarks upon experience. It does not indeed follow that a leaf which has been observed to produce roots will be suitable for propagating; for we know many hard-leaved plants the leaves of which readily produce adventitious roots and remain fresh for months, but which will not form adventitious buds.

Begonias are perhaps most often propagated by leaf-cuttings, and in their case it is chiefly the so-called leafy Begonias, like

Begonia Rex and its varieties. The strongest ribs are cut through and the leaves are laid with their under surface on damp sand. If care be taken to prevent the leaf from drying up by keeping the atmosphere saturated, and if the temperature is sufficiently high, it will be noticed that the severed ribs are very soon covered in with callus. The wounds become surrounded by an excrescence of tissue which at first produces only roots. Very soon, however, adventitious buds make their appearance on this tissue. These buds, however, do not produce their own roots, but content themselves with those formed on the callus.

It is especially interesting that these young shoots arise from one, or only very few, epidermal and subepidermal cells in the neighbourhood of the severed vein; thus we see that even tissues with very little protoplasm are able to enter again into cell-division if they are well supplied with reserve materials. The new roots take their origin much deeper, generally in immediate proximity to the cambium layer of the vascular bundle, which runs along the vein. But even before the roots are developed the leaf seeks to develop absorptive organs, and we find the epidermal cells near the wound growing out into hair-like processes, which somewhat resemble root-hairs.

As far as our observations go, we may assume that the formation of young plants from leaf-cuttings takes place in other plants in much the same way as in the case of Begonia. Some differences occur, as in some cases (*Peperomia*) no callus is formed over the wound, but the latter is simply protected by a layer of cork cells. In other cases the young buds are not formed directly from the tissues of the leaf, but arise from the callus tissue (*Achimenes*). Such differences are, however, unimportant, and may occur in the same species under different conditions of nutrition.

Among the Monocotyledons it is especially the bulbous plants, the leaves of which are prone to form buds, and can therefore be used as cuttings. It is well known that many liliaceous plants may be propagated by bulb scales; but it is not generally known that, if properly treated, the green leaves will produce new bulbs. The young buds which arise on the cut end of a bulb scale or of a green leaf originate from the epidermal or (in older leaves) from the subepidermal cells, their appearance being preceded by the development of a certain amount of true callus, the outermost cells of which are the actively dividing ones. When the small tubercle of callus tissue has attained a certain size, then a ring-like protuberance makes its appearance, which grows up in the form of a sheath round the apex and forms the first leaf or scale of the young bulb.

The formation of these small bulbs (*bulbils*) is best seen in the case of *Lilium auratum* or *tigrinum*. Here they make their appearance on the inner surface of the leaf near the margin, while the roots arise from the bast region of the fibro-vascular bundle. The latter are, however, only of very short duration, as the new bulbs very soon form their own roots.

In most cases, however, it is not necessary to stimulate the leaf to the formation of buds by wounding it, as the leaves will readily form buds of their own accord. This voluntary formation of buds is not in any way different from the bud formation of cuttings. Examples of it may be found in the groups of Mosses and Ferns, as well as among the Lilies in the group of Monocotyledons. It occurs frequently, too, among Dicotyledons, where the buds are generally formed in the axils of the veins, and are larger and stronger the larger the vascular bundle is next which they lie.

It is very well known that the leaves of *Bryophyllum calycinum* produce such buds in the depressions between the teeth of the leaf, where a meristematic tissue occurs, which soon begins to form buds.

As such buds, when carefully removed, may readily be grown into separate plants, the knowledge of plants producing such bulbils is useful to gardeners. We will, therefore, give a list of plants in which they will be found to occur:—*Hyacinthus Pauzolsii, Fritillaria imperialis, Atherurus ternatus, Ornithogalum thyrsoides, Drimia, Malaxis, Cardamine, Nasturtium, Tellima, Siegesbeckia, Utricularia, Calanchoe, Begonia quadricolor* and *phyllomaniaca, Nymphæa micrantha* (according to Caspary), and its hybrids. If the gardener desires to use the leaf-buds of these forms for propagating, it is advisable to keep the plants from flowering, and to place them in a moist and shady situation. The vegetative activity will thereby be increased and more bulbils will be formed.

CHAPTER X

THE THEORY OF WATERING

§ 40. Why must we pay special attention to the watering of plants?

AFTER having examined the structure and function of the stem and leaves, we must consider the conditions which conduce to a normal development of these organs. We have already dealt with the subject of transplanting, and assuming that the roots will find a requisite amount of food material in the soil, we must consider how the supply of water should be regulated to ensure the continued activity of all the organs of the plant.

It is a fact that the older a gardener grows the more care he takes with the watering of his plants; for year by year his experience teaches him more definitely that careless watering is the cause of an uncommonly large number of diseases of plants.

The difficulty in watering lies in the fact that plants require various amounts of water according to their species, their age, their situation, the season of the year, and their actual state of health or development. The amount of water which is sufficient at any given time may be very much too large a month later, and may therefore cause considerable injury. Above all things, we must remember that the **transpiration from the leaf surface is not a merely mechanical, but a physiological process,** and as such is regulated by the activity of different internal organs, just as much in the case of plants as in that of animals.

It is, of course, true that the conditions which increase the transpiration of an inanimate object in general accelerate the transpiration of plants, but this is by no means always the case.

The fact that the transpiration of living plants varies generally on the same lines as the evaporation of water from an inanimate body will be understood if we reflect that the factors which increase the evaporation in the latter case (especially the rise of temperature) increase the vital activity of the vegetable organism. An increase of the vital processes, however, entails an increase of transpiration. On the other hand, some conditions may occur under which the amount of water given off by plants decreases, while inanimate bodies would show an increase of transpiration. Thus sickly plants may show a diminution of transpiration with a rising temperature. Inversely, however, an increase of transpiration will be manifest even with a fall of temperature if plants are transported from a strongly concentrated nutritive solution to a more dilute one. The roots will then appear to put forward their best energies to absorb from the weaker solution as much of the inorganic food matter as they formerly obtained from the more concentrated solution. But this is only possible when in the same interval of time larger quantities of the water are taken up, which means, of course, that a greater amount must be given off too.

The fact that sickly plants are not able to give off as much water as healthy ones must not be forgotten when plants with injured roots are re-potted and placed in a hot-bed. It is then often supposed that the increased temperature of soil and air will stimulate the plant to increased transpiration. The pot is therefore often kept continuously wet, and the decay of the roots, against which one is fighting, will continue unabated.

The amount of water necessary for plants is determined—

(1.) **By the kind of plant under consideration.**

(2.) **By the intensity of the vital processes at any given moment.**

To know the requirements of any special kind of plant, it is necessary to know something of the conditions under which the plant grows in its native country. As far as practical experience has taught us at present, we may assume that plants from rocky habitats, without shade, and from countries with long periods of drought, are able to do with a very small supply of water, and can only put up with an occasional drenching. In those cases in which the climatic condition of the native country are unknown to us, the external appearance of the plant may give us some clue to its proper treatment. Plants with narrow and tough leaves, especially when the leaf-blade is vertically placed, do not, as a rule, like much water; plants with broad leathery leaves prefer a damp atmosphere to great moisture of the roots. Succulent plants with hard epidermal cells (leafless Euphorbias, succulent Compositæ, Aloes, and Agaves), and thin-leaved plants with a strong woolly covering of hairs, are further examples of plants which require little water.

Then the structure of the roots may often reveal the requirements of the plants. Succulent roots in the case of thin-leaved plants, and less succulent ones which taper away and grow rapidly to the bottom of the pot, usually like, and often require, a large amount of water. On the other hand, a multiplicity of fine branching rootlets which tend to grow out above the surface of the soil can often do with a small amount of water, but require a greater amount of aëration.

In watering we must attend quite as much to this desire of the roots for air as to that for water.

As we have already had occasion to mention, the roots require just as much as the stem structures a thorough supply of air. If this is not forthcoming, alcohol is formed in the cortical tissue of the root, and later on fusel oil and acids (acetic acid, butyric acid, &c.) make their appearance, and these acting poisonously, generally cause the death of the organ. The products of decay are then transported from these dead portions of the plant to the still living organs, and the inevitable result is the death of the latter.

The most frequent cause of the formation of alcohol in roots and the subsequent decay of the latter is a permanently wet ball of roots. The unfounded fear of damage arising from a lack of water causes gardeners to water the pots as soon as the surface of the soil begins to dry; and it is not taken into consideration that the centre of the pot still contains a large amount of water. The consequence is that the lower part of the pot is continuously saturated with moisture. Thus the air is prevented from passing freely to the roots, and these are gradually suffocated. This suffocation of plants takes place very slowly, and is preceded by a long death-like rigor, which is only overcome by a renewed supply of oxygen. If death were to ensue rapidly in these cases, we should see even a greater number of pot plants die than are actually killed at present by too much watering.

We need only observe the tremendous variations in the amount of water supply to which these plants are exposed, when they are dependent upon the rain, to see how little harm these variations do. We are, therefore, generally in danger of injuring the plants by a too anxious endeavour to supply every small amount of water lost by the plant. Indeed, it is very good for plants, just as for human beings, to feel the passing pangs of hunger or thirst, and they take no harm if they are not watered until **the pot sounds hollow.** If water is then supplied, it should be done so copiously that it runs out of the hole at the bottom of the pot. The practice of inexperienced persons of repeatedly wetting the upper surface of the soil (especially of plants in peaty soil) has the disadvantage that the lower portion of the soil remains dry while the surface and the base of the stem become covered with moss. The above rule is especially important in the case of tough-leaved plants, which do not require much moisture. It is different in the case of herbaceous plants which are rapidly growing. In their case every diminution of the water supply is injurious, and in the case of these rapidly-growing plants there is very little danger that the soil will be water-logged for any considerable time, and consequently stop the respiration of the roots.

But even if we know the natural conditions of the plant and the requirements of the species, we will not be able to water the plant rationally if we do not know or do not consider the peculiar requirements of every individual. Even the same individual with the same amount of leafy tissue may need a larger or a smaller amount of water according to the vital activity which it manifests at any given moment.

This difference in the amount of water required at different periods becomes more comprehensible if we remember that the water is not only needed as a food substance, the elements of which become separated in the vegetable tissues, and used up in the formation of new organic substance, but that it serves also to transport the mineral salts contained in the soil to the plant. These are used up by the plants more rapidly at one time than at another, and the plant will, therefore, sometimes take up more of the solution, sometimes less.

From experiments made on the transpiration of plants, we know that every plant draws most largely on the soil during the putting forth of new shoots and leaves, and takes up most mineral substances at this time. At this period, even plants which grow in arid regions like and require considerable supplies of water. As soon, however, as the shoot is developed and the formation of flower-buds is expected, the watering may be decreased.

The formation of flower-buds is best initiated by preserving a period of rest, and the latter is favoured by a diminished water supply.

This is the reason why cactus-growers let their plants almost shrivel up after the shoots have been produced, and for the same reason pine-apples receive less water when the formation of fruit is desired. If water is freely given, and no resting period is allowed, leaf development alone takes place, as the tip of the shoot grows on continuously, and remains the centre of attraction for the ascending sap.

This condition is not desired by the grower, but does not injure the plant. In other cases, however, untimely watering may produce actual disease. This may occur, for instance, when **the period of rest** at the end of the vegetative season **has been disregarded.**

The activity of a plant during one vegetative season may be divided into two periods, and this is the case whether the vegetative season is our summer, or, in the case of tropical plants, falls in with our autumn and winter. With the awakening activity, the first purely vegetative period commences, during which the shoot is produced, and begins during the first few months to accumulate food material. In the case of annuals, the organic material which has been formed in the leaves is immediately used up during the flowering period, which follows upon the period of leaf expansion. The food material is immediately and continuously withdrawn from the leaves, and the latter begin to fade away, beginning near the base of the stem, and die. In perennial plants the material which has been accumulating during the first period is used during the second, either in the production of a later shoot (summer shoot) or for the formation of flower-buds which will open in the subsequent spring. In the case of our trees, a great part of the nutritive substance is conducted down from the leaves and stored up in the base of the stem and the main roots, and gradually, when these organs are filled, in more peripheral structures, the lateral branches and the rootlets. This process of storing food material comes to an end when the fall of the leaves ushers in the period of rest. As this partly depends upon the gradual decrease of temperature, the upper portions of the stem, which are exposed to the atmosphere, may have completely entered the period of rest, while the roots, which lie in the warmer layers of soil, are still continuing their growth in thickness. This activity may sometimes last until January, and then only can we say that the plant is entirely at rest.

Such a period of rest occurs in all plants which last for more than one year. It is either the cold, which stops all vital processes, or, as in tropical regions, the excessive drought of the hot season. The rest due to this latter cause is so important that tropical evergreen parasites belonging to the group of Phanerogams will lose their leaves when they grow on trees which shed their leaves during the dry season. A very excellent example of this phenomenon is *Loranthus longiflorus* when growing upon *Sterculia, Spondias, Terminalia*, or *Erythrina*.

Such a period of rest cannot be interrupted with impunity, but it can be shortened by artificial means. Nature itself allows the resting period to be of different durations, according to the development attained during the preceding period of activity. This is best seen in the forcing of bulbs, the latter being able to develop very early if the preceding month of May was warm and dry, so that the bulbs could ripen well and early.

During this latent period plants require very little nourishment, and even those green-house plants which retain their green leaves require an infinitesimal amount of water and of nutritive salts.

But how often is this experience disregarded. In the coldhouses in which *Erica, Epacris, Rhododendron, Azalea, Acacia, Melaleuca*, and other Australian and Cape plants are placed, you may see the careful gardener watering every day those plants which show the slightest sign of dryness on the surface. The consequences very soon show themselves. The stems of the Ericaceæ begin to rot, Leguminosæ and Myrtaceæ lose their leaves, and other plants lose their flower-buds. Decay of the roots sets in in all cases.

The disregard of the resting period makes itself felt in other ways too. Many growers believe that by increasing the temperature and moisture they can awaken any resting organ, and by trying to do so they often come to grief in the case of tuberous plants. If the tubers of *Caladia, Cyclamen, Begonia, Gloxinia*, &c., are placed in a hot frame for forcing, are perhaps covered up with soil and kept permanently damp, it is not germination but decay that will set in. It is generally the mould (*Botrytis cana*) which causes the destruction of the tubers. In some cases they can be saved by cutting out the rotten portion, covering the cut with powdered charcoal, and placing the tubers on the surface of the soil, where they will have more light and more air.

It must be remembered that, with few exceptions, the transformation of starch into soluble sugar can only proceed slowly in the storage tissues. It is also only possible to gradually induce this change, and in the first instance it takes place in bulbs and tubers without much water. This we may readily conclude from the fact that such succulent organs begin to sprout at the proper time, even when they are not buried in the ground and receive no liquid water.

The calling of a dormant organ into life or activity must take place gradually, and during the first development the supply of water must be but small.

Lastly, there are still some symptoms to be mentioned which should warn gardeners to be careful in the watering of their plants. They make their appearance when a damp atmosphere accompanies a wet soil. For though a damp atmosphere is useful and necessary at the time of active assimilation, it can be very detrimental when it is excessive during the period of rest. A saturation of the air with water vapour, especially if the temperature is low, is favourable to the activity of Fungi of decay in their attack upon plants. The activity of the latter is decreased by the moisture of the atmosphere, as assimilation and transpiration do not then take place to any great extent. For the fungal spores, however, this temperature is sufficient, and their development is accelerated by the great moisture.

If the temperature of green-houses is raised in the winter when illumination and aëration is insufficient, then the plants, stimulated by the moisture to further growth, may develop along wrong lines. The different processes which take place in a leaf make different demands upon the different factors, such as light and heat. Thus the elongation of cells can take place in feebler light than can the assimilation performed by the leaves. The formation of sugar can take place at temperatures at which it is no longer used up in the process of respiration; sugar may therefore often be stored up at low temperatures in leaves, tubers, and other organs.

It often happens also that in green-houses the temperature is high enough to permit of an elongation of the cells, but the light may not be sufficiently intense to cause any appreciable assimilation. If under these conditions the pots are copiously watered, while only a small amount of evaporation takes place from the leaf surface, a considerable accumulation of water will occur within the plant, and this will cause a pathological (abnormal) elongation of the cells. Some groups of cells within the tissues of the leaf will begin to elongate, and even grow out into long tubes, but as they receive no material from without for the elongation of their cells, they use up their own cell-contents. Thus the chlorophyll corpuscles become disorganised, and nothing but small yellow granules will be found. This causes such impoverished portions of the leaf to appear more or less yellowish, even by reflected light, or at all events when the leaf is held up to the light.

Such appearances are of much more common occurrence than was formerly thought, and it must now be regarded as a sign of a superfluity of water.

In other cases the discoloration is less intense, but the elongation of the cells is so great that the epidermis is lifted up into small wart-like protrusions on both the upper and on the under surface of the leaf.

The formation of yellow patches on leaves as described above occurs in the leaves of *Pandanus* and *Dracœna*; the wart-like protuberances of the leaf surface may be observed in *Ficus elastica*, *Aralia*, *Hedera*, *Solanum*, and other plants.

The appearance of yellow patches with indistinct margins or of small glandular dilatations on leaves during the resting period may be taken as signs that the plants are probably suffering from a superfluity of water, and the watering should be decreased.

CHAPTER XI

THE FLOWER[1]

§ 41. Of what parts does the flower consist?

In the case of a complete flower we must distinguish two groups of organs. We have first of all the essential organs of reproduction, the stamens and the pistil, and secondly the less important protective organs (sepals and petals), which may be only rudimentary, or indeed entirely absent. If there is only one set of protective leaves, the covering formed by them is termed the **perianth**. The simplest flowers may consist of a single sexual organ, as, for instance, is the case in the female flowers of the Yew (*Taxus*). Usually, however, a number of different organs are collected in a flower, and are often arranged in a number of whorls one below the other, and then each group of such organs has received a special name. All the male reproductive organs (*stamens*) are collectively termed the **androeceum**, while the female reproductive organs (*carpels*) form the **gynæceum**. If the male and female organs are surrounded by the same perianth, the flower is termed **hermaphrodite**, while a unisexual (*diclinous*) flower will contain either stamens or pistil only. The separation of sexes may be of such a nature that both occur at least on the same plant (*monœcious*), or the separation is so thoroughgoing that each individual has only sexual organs of one kind (*diœcious*).

---

[1] At the commencement of this chapter we again point out that the reader must not expect a description of the various modifications of flowers occurring in the different natural orders of flowering plants. This book deals with the functions of the flowers as a whole, and we shall discuss them in the case of a flower in which they are most typically represented. The systematic description of the peculiarities of the different families, orders, and genera will be found in every text-book of botany, and we only mention here those anatomical and physiological peculiarities which are necessary for a scientific application of horticultural methods.

In most cases we are justified in regarding the reproductive organs as leaves modified for a very definite object, and the stamens have therefore been termed **staminal leaves**, while the pistil may be looked upon as built up of one or more carpellary leaves or carpels. The most important part of the stamen are the pollen sacs, which run parallel on either side of the median connective. Looking upon it as a modified leaf structure, the stamen represents a leaf consisting of a long and delicate stalk (*filament*), and thick cylindrical blade (*anther*), into which the stalk is continued as a connective, the top of which often projects above. On both sides of this **connective** lie the pollen sacs, two of which go to form each half of the anther. The stamen has therefore four pollen sacs. The tissues in the middle of each pollen sac will undergo certain divisions, and the cells which ultimately result from these divisions round themselves off one from another. They become covered by a thick outer coat, which is often curiously sculptured, and each cell is termed a **pollen grain**. These are generally liberated by the longitudinal splitting of the anther, which opens the pollen sacs. In some cases the anthers open by pores (*Ericaceæ*).

The stamens have, therefore, in many respects, the same structure as the normal green leaf, and we must, therefore, not be surprised if at times, when there is a considerable tendency for the development of green leaves, the stamens themselves should become transformed into green leaves.

The anthers in that case do not become cylindrical, but are thin and expanded, and coloured green. The loss of the cylindrical shape of the anther generally entails the loss of the power of forming pollen, and the stamen becomes a true green leaf. This **retrogressive transformation** is at the bottom of most doubling of flowers.

The female organ or pistil can be very well seen as the central structure of the flower of the Tulip or the Cherry. It consists of a dilated basal portion, **the ovary**, a long tubelike prolongation, **the style**, with a knob-like or lip-like termination, the **stigma**. We may conceive it to be formed by a number of sessile leaves (*carpels*), the blades of which are drawn out into points, and the margins of which have become fused together. Of the three portions of the pistil, the lowest one, the ovary, is the most important, as it contains the ovules with their egg-cells. The stigma, which in some plants is very extensive (feathery in Grasses, disk-shaped in the Poppy), may be looked upon as an organ for catching the pollen, while the style conducts the pollen tubes, which grow out from the pollen grains, down to the ovules. In many cases there is no style formed, and the stigmatic surface is directly on the top of the ovary.

The sexual organs are, in most flowers, surrounded by the leaves of the floral envelopes, which are either arranged in several whorls or in a spiral manner. In complete flowers the floral envelopes are sharply differentiated by form and colour into two whorls of leaves, the inner one of which, the **corolla**, is built up of delicate cells, filled with a colourless or a brilliantly coloured cell sap. The epidermal cells of these leaves are often drawn out into papillæ, and give the surface of the petals a velvety appearance. The outermost whorl of leaves, the **calyx**, is generally green, and resembles greatly in texture, and often also in shape, the assimilating leaves.

If we cut through a flower longitudinally, we see that the several whorls of leaves appear to spring from one point. In reality they are, however, separated by short but distinct internodes of the axis.

Each flower represents, therefore, a reduced shoot with several whorls of leaves, each of which is modified for a special function, but which, under certain conditions of nutrition, exhibit the tendency to assume the nature of a green assimilating leaf.

Such a retrogression of the flowering axis to a leafy shoot we term **proliferation.**

The flowers occurring in some of the natural orders, however, diverge often very considerably from the typical flower described above. One of the most frequent modifications results from the disappearance of the differentiation of calyx and corolla. The floral envelopes are completely alike, and either resemble a corolla (Tulips and other Liliaceous plants), or are green and have the appearance of a calyx (*Juncaceæ* and *Cupulifera*). In such cases we speak of a **perianth.**

Sometimes the sepals and petals may differ one from the other in their appearance, but the leaves composing these two envelopes are spirally arranged, and gradually pass over from one condition into the other, as is the case in the Water-lily (*Nymphæa*), the Queen of the Night (*Cereus*), and the All-spice (*Calycanthus*).

Sometimes the petals are much reduced or transformed into special organs for attracting insects by the secretion of honey, and are then termed **nectaries**. In either case the calyx will then undertake the functions of the corolla, and will become brightly coloured, as, for instance, in the Larkspur (*Delphinium*) and the Monkshood (*Aconitum*).

In some cases we find the food material available for the development of the flowers is differently utilised in different flowers of the same species. The female flowers, which require a large amount of nitrogenous food material for the formation and development of their ovules, only produce a very small and inconspicuous corolla, while the male flowers use the food material for the development of a comparatively large and corolla-like perianth, as may be seen in the case of the Hemp.

On the other hand, we find occasionally a tendency in some plants to produce a more elaborate corolla than is normally formed. In such cases, at the point where the petal or the perianth-leaf narrows down into a stalk, there will be produced on the inner surface of the corolla outgrowths of the same texture and colour as the corolla, which form what is termed the **paracorolla** or **corona**. The Pheasant's-eye Narcissus possesses a structure of this nature of a delicate red colour, while in the Daffodil or Lent-lily it is larger, bell-shaped, and more like the corolla in colour.

The thousands of variations which may occur in the form and colour of flowers do not change the general conception of a flower which we have put forward, namely, that the flower is a modified shoot, the different leaves of which may become transformed from one series of forms into another. It is the business of the horticulturist to change the nature of the floral leaves to suit his special purposes.

§ 42. How are single and how are double flowers developed?

If we examine the most perfect type of flower that is, one provided with a calyx and corolla, we find the modified shoot, which we term a flower, begins its development as a slender meristematic papilla, at the base of which a number of small ring-like outgrowths appear. The papilla, which first makes its appearance, is the axis of the flower, and the first outgrowths from its young cortex (*periblem*) are the calyx leaves. The first leaf rudiments grow more rapidly on the outside than on the inside, and thus form concave structures which arch over the apex of the axis and protect it. Protected by these leaves, another series of outgrowths appear farther up the axis, agreeing in arrangement and number with the petals. They are indeed the rudiments of the petals, but they behave differently from the sepals in their development, remaining rudimentary and scale-like for a long time, even when the new outgrowths arising above them have been transformed into stamens and carpels. It is only when the flower approaches the period of expansion that the petals grow very rapidly, become coloured, and by their increase in size burst open the calyx.

According to this representation, therefore, the flower must be conceived of as a shoot at the base of which the sepals are formed, and which produces nearer its apex the other floral leaves, the petals and stamens, while quite at the apex the carpellary leaves arise and form the pistil. If we conceive the axis which bears these floral organs to be made of a plastic substance such as clay or putty, and imagine the apex of the axis introverted, so as to form a cup-like structure, we should find the pistil in the centre of this hollow receptacle, and the other leaves higher up on the margin. This is the case in the so-called perigynous and epigynous flowers, of which Fig. 30 represents one in longitudinal section. It is a young apple-blossom.

In this flower the actual apex of the shoot, which in the previously described (hypogynous) flower occupied the summit of the flower, is found at the base of the cup, and the cells constituting it are shown in Fig. 30. The margin of the cup is made up of the sepals (*r*), from the base of which springs, as shown on the left-hand side, a tongue-like projection directed upwards, one of the petals. Inserted at the same point is one of the many stamens with only a very short filament. Lower down are seen two lyre-like arms, which are the styles of two of the pistils, the lower portion of which is cut into on the right-hand side and exhibits the ovarian cavity, in which the seeds would later on be contained. The outer portion of the section up to the point *f* consists of the hollowed axis or receptacle, which during the ripening becomes succulent and represents the edible portion of the apple. We see, therefore, that the apple is really the cortical tissue of a succulent shoot.

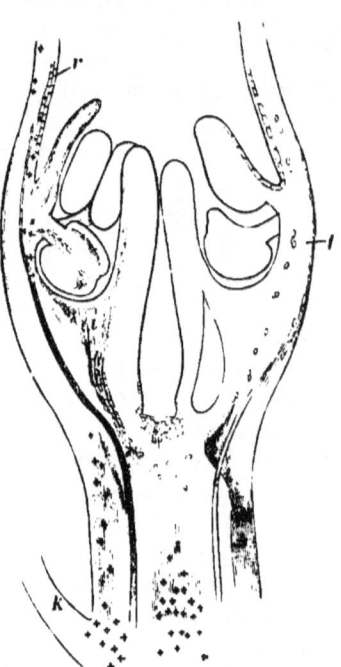

FIG. 30.—YOUNG STAGE IN THE DEVELOPMENT OF THE FLOWER OF THE APPLE.

It is, therefore, not startling to find that in some varieties of pears the sweet and succulent tissue extends some way down the fruit-stalk. The process of forming succulent cortical tissue has only extended a little farther downwards in this case.

We can now also understand how in some cases pears are formed which have no core: they are derived from flowers on which no pistil was developed, but in which the axis followed the normal course of development.

In those cases in which the petals begin to elongate most rapidly just before the flower opens, while the growth of the calyx has been more rapid up to that period, the corolla functions usually only as an attractive organ. If, however, the calyx is only small, and the corolla undertakes the function of protecting the sexual organs, as is the case in the flowers of the Vine, then it is developed as rapidly as the calyx and precedes in its development all the other organs.

On the other hand, it is rare for the calyx to undertake the functions of the corolla. Instances of this have been mentioned in the case of *Aconitum* and *Delphinium*. In some rare cases the doubling of flowers depends upon this. Thus double Primroses (*Primula veris*), Canterbury Bells (*Campanula*), and *Mimulus* become double by the transformation of the sepals into petaloid leaves.

But it most frequently is the case in doubling flowers that the calyx remains unaltered and that the number of petals increases very considerably. This may take place by the transformation of stamens into petals, as it occurs, for instance, in the Ranunculaceæ. The tendency to produce leaf-like organs may, however, be so great that each stamenal rudiment divides into several pieces, and each stamen will therefore be represented by several petaloid leaves (*Caryophyllaceæ*). Sometimes the stamens remain unchanged, and then the rudiments of the petals are formed of such a breadth that these leaves divide up into several leaves, as happens often in the Fuchsia. When the flowers of Clarkia become doubled, an actual proliferation takes place at the base of the petaloid leaves formed from the stamens. In some cases indeed such proliferation may occur on the axis itself between the staminal whorl and the petals (*Campanula*).

In the case of Composites, the process of doubling is a different one. What we generally call a flower is in this case a basket-like head of flowers, *i.e.*, a number of small distinct florets, which with insufficient examination we should consider as petals, are inserted on a cushion-like receptacle. The florets may also be of two distinct kinds, as, for instance, in the Cineraria, where the outer ones are brilliantly coloured, the tube of the corolla being drawn out into a long tongue-like structure formed by the fusion of the several petals (*ray-florets*). The centre of the receptacle is occupied by inconspicuous tubular florets (*disk-florets*.) The so-called doubling of such flowers may be brought about by the transformation of the disk-florets into florets similar to the ray-florets, which often, however, is accompanied by a deficient development of the sexual organs. Similar transformations take place in the production of the double Guelder Rose (*Viburnum Opulus*) and in *Hydrangea*. In the latter case, it is, however, the calyx which is the conspicuously developed portion of the flower.

In the case of the capitulum of the Composites, the little basket which encloses the flower, and which is often falsely termed the calyx, is formed of a number of overlapping scales, and should rightly be called the **involucre**. Besides these, a number of bracts may be found between the several florets and subtending them. Both these kinds of bracts may be considerably enlarged and conspicuously coloured, and the capitulum then has the appearance of a double flower. This is the case with the "Everlasting Flowers" (*Xeranthemum, Helichrysum, Acroclinium roseum*, and *Rhodanthe Manglesii*). Such capitula very rarely divide, but occasionally small lateral capitula make their appearance between the bracts of the involucre.

The tendency to "double" is always less in the case of plants with a corolla in form of a single tube (*gamopetalous*) than when the corolla consists of a number of separate petals (*polypetalous*); but among the latter there are some natural orders which have no tendency to "double" (*Umbelliferæ*). Speaking in general terms, we may say that it is easier to produce double flowers in plants with radially symmetrical than with bilaterally symmetrical flowers.

Double varieties have, however, been produced in the case of *Viola, Pelargonium, Impatiens, Pisum, Azalea, Lobelia, Gloxinia,* &c.

§ 43. Can a gardener determine the development of flowers ?

This question really resolves itself into two questions:—1. Can a gardener cause the formation of new flowering buds? and, 2. Can he cause a change in the course of development of existing floral rudiments? In answering these questions, we cannot at present base our answers on precise experiments, but we must make use of general experience, which has been accumulated during years, or indeed centuries, of horticultural practice. Science has yet to make many investigations into such practical subjects.

In all cases in which the horticultural value of a cultivated plant lies in its flowers, or in the resulting product of those flowers, the fruit, it is desirable to increase as much as possible the number of flowering buds. On this point we know by experience that **plants will only develop flowering buds when the food material formed in the leaves is copiously stored up in the stem and branches as reserve material, and not when this material is immediately used up in the production of new vegetative organs (leaves).**

In horticultural practice it is common to observe in the case of perennial plants that a continuous and excessive formation of leaves is detrimental to the production of flowers (pine-apple and vine growing). Of our apple-trees it is well known that in warm insular climates they grow into magnificent foliage trees, but remain unproductive of fruit. Growers of cactuses will have found out that if the plants are copiously watered during the winter in a warm house or room, they will very rapidly produce new shoots, but no flowers. Many Australian plants (*Metrosideros, Cantua dependens, Correa*) are generally kept in small pots by gardeners because they are said to blossom better.

In all these cases it proves to be advantageous to prevent the development of shoots and to bring about a complete period of rest. The best means of doing this is by drought, and under certain circumstances by diminution of the temperature. We have already touched upon another method of attaining the retention and storing up of food material by ringing and notching the branches.

That a diminution of the supply of water accompanies the production of flowers in nature may be gathered from the fact that most trees and shrubs produce their flowers on short reduced shoots or spurs. The comparison of the anatomical structure of such a short shoot with that of a long leafy shoot confirms our statement, too, that an increase in stored food material is necessary for the production of flowers. The former shoots have far more storage tissue than the latter; the development of the parenchymatous pith and cortex as compared with the woody cylinder being very much greater in the former. This is most conspicuously the case in the development of the so-called "pouches" in some varieties of pears, in which the spur is very short and almost succulent, and has only half as much woody tissue as a leafy shoot of the same age.

In forcing shrubs and trees, therefore, it is quite a rational procedure to place the plants in pots quite a year before the forcing begins, and to nourish them very plentifully, but to begin to reduce the water supply towards the end of the summer. In consequence of the gradually occurring drought, the autumn rest and fall of leaves occurs at an earlier period, and the changes which take place in the cells during the resting period are as far advanced at the commencement of the winter as they would be under ordinary circumstances in the next spring. The withholding of water in such a treatment prevents the use of the assimilated food substance for the growth of new shoots and causes it to be stored up near the buds.

The success in the cultivation of bulbs also depends upon a period of dryness occurring at the proper time. A damp spring will cause the production of strong foliage in a well-manured soil, and an early, dry, and warm summer following upon this will prevent the leaves from growing too long, and will cause the production of a large number of flowering buds. If the summer continues dry, so that the leaves fade soon and the bulbs ripen well, then we shall have richly flowering hyacinths, tulips, &c., which will lend themselves to forcing. The very copious manuring which has more recently been applied to the growing and flowering bulbs does nothing more than cause a luxuriant development of the flowers which are already formed, but will not cause the production of a greater number. Weak flowering buds, which, if badly nourished, would not open, will freely expand if the plants are well manured.

The use of manures will be beneficial for the solution of the problem which horticulturists endeavour to solve, namely, the production of changes in the various organs of the flower. The desired change is generally the increase of the petals in size and number at the expense of the sexual organs of the flower.

All the processes which cause a doubling of flowers follow one line of development, and that is a reduction of specialised organs to their primitive leaf-like condition and the development of new whorls of leaves in the corolla.

As we have already stated in speaking of the retrogressive changes which may take place in flowers, all the different organs of the flower may become transformed into green leaves. If we desire, therefore, a production of foliar organs in the flower, we must use the means which are favourable to the growth of leaves. The latter depends firstly upon the supply of food material, and secondly upon the amount of water available for its requirements. If there is a plentiful supply of food material, especially of a nitrogenous nature, the multiplication of the cells will be a very active one. If at the same time the organs are continuously and richly supplied with water, the young cells will also be able to grow to their utmost capacity. While for the production of flowering buds, it is essential to decrease the supply of water and of nitrogenous salts, to increase the phosphates supplied to the plants, and to increase the illumination, double flowers require for their development copious watering, an increase in the nitrogenous salts in the soil, and, according to our experience, a decrease in illumination. These factors, either singly or combined, may be observed to have taken place where a leafy development of flowers has taken place naturally, *i.e.*, without the aid or intention of man. On the other hand, we often find that faults in the treatment of plants may produce a leafy development neither desired by the grower nor desirable. The proliferation of the catkins of the Hop, which is due to the elongation of their axis and to the abnormal development of leaves upon the latter, is caused by excessive moisture or too rich a manuring with dung, or even with Chili saltpetre.

A change in the vegetative periods, too, has been shown to produce doubling. In the case of the common Cineraria

(*Pericallis cruenta*), weakly specimens which have not completely developed their inflorescence in the spring, and are then bedded out in a shady place, will show in the course of the summer some curious appearances of doubling. The connectives of the stamens enlarge above the anthers and become coloured like the petals, and the ray-florets develop excrescences which give them the appearance as if two ligulate flowers were fixed back to back. Sometimes, too, adventitious heads of flowers become developed in the axils of some of the bracts of the involucre. These malformations are brought about because the development of the flower takes place not during the light and cool spring weather, but in shady places with a considerable amount of moisture and heat during the summer.

The effect of rich manuring in transforming flowers is best seen in the case of tuberous Begonias. Here it is not only the male flowers which have been doubled by the transformations of the stamens into petals, but the female flowers too have become transformed in a similar way. These flowers, too, lend themselves very well for the study of such transformation. The branches of the style increase in number and expand like leaves. Indeed, the transformation may be more far-reaching still and extend down to the ovary. In this case the ovary will open, and the ovules will protrude into the flower like small white grains. Indeed, when the style and the placentæ themselves are transformed into leaves, the ovary will have seemed to have disappeared, and the ovules will cover the base of the petals as a fine white powder. The ovules themselves under the microscope will often show a partial transformation into leaf-like organs.

Such examples demonstrate sufficiently well the lines along which the gardener has to work to secure changes in the development of the various floral organs.

CHAPTER XII

FRUITS AND SEEDS

§ 44. How are fruits and seeds formed?

IN discussing the structure of a complete flower, we mentioned that the centre of each flower was occupied by the pistil, which consists of an ovary, a style, and a stigma. Of these three parts the stigma is the apparatus for catching and retaining the pollen, the style the channel along which the pollen tube grows when the pollen grain germinates, and the ovary is the receptacle where the ovules are produced and develop into the seeds. The pistil is either formed of a single carpellary leaf, the margins of which have been folded in and have become fused, or several leaves arranged in a ring have become fused together, in which case usually the ovary will consist of several chambers. As already mentioned, the tube-like elongation between the ovary and the stigma, the style, is relatively of little importance, and is not developed in some plants, the stigma being situated directly on the ovary. Such is the case in the pistil of the grass represented in Fig. 31. Here the stigma is feathery and very delicately developed; its numerous filamentous branches rendering it very efficient in capturing the pollen grains, which are easily carried about by the wind. The main branches (g) which bear the feathery processes are directly continuous with the tissues of the ovary (f), which presents only a single cavity, the latter being completely filled by the single ovule (o). This consists of a very delicate tissue, which is differentiated into a central portion and an enveloping layer. This latter is formed by two **integuments** (i), which do not completely close in the central **nucellus**, but leave a very small passage open to the latter. This opening is termed the **micropyle**.

FRUITS AND SEEDS 227

If some pollen grains chance to adhere to the feathery stigma, they develop a long tubular process by the protrusion

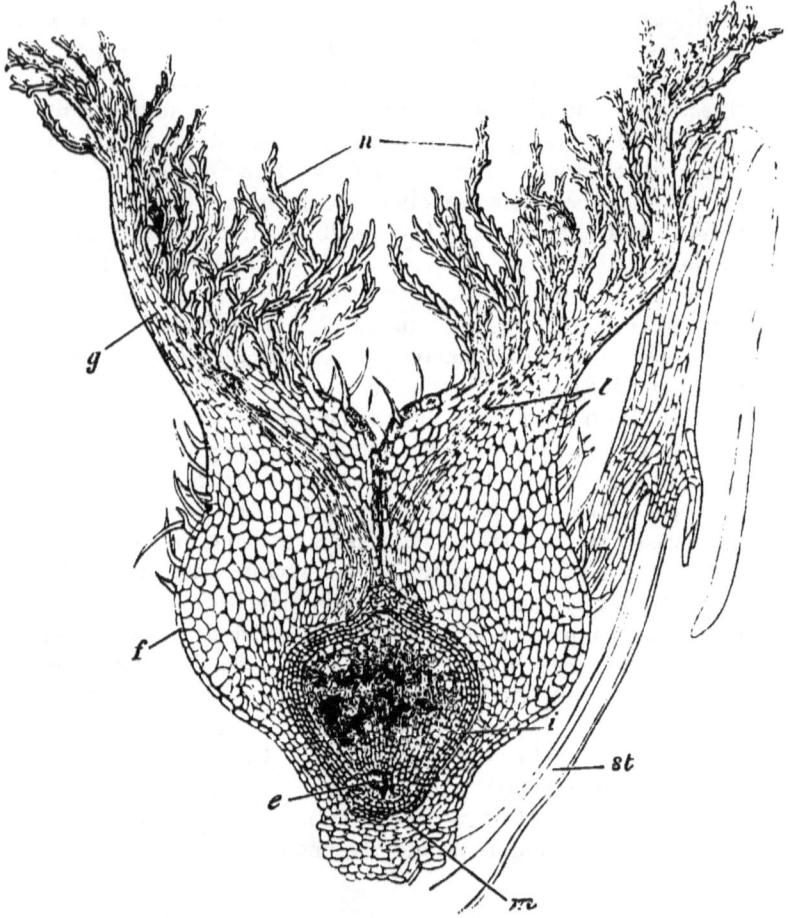

FIG. 31.—YOUNG PISTIL OF THE BARLEY.

st, stamen; *f*, ovary; *n*, stigma with feathery branches; *l*, the tissue through which the pollen tube grows; *o*, ovule; *i*, the integuments; *m*, micropyle; *e*, embryo sac.

of the delicate inner coating (**intine**) through the thin portions of the hard outer coat (**extine**). This pollen tube makes its way down through a special conducting tissue (*l*) into the cavity of the ovary, and then through the micropyle (*m*) of the ovule. The more detailed representation of these processes is shown in the subsequent figure (Fig. 32). In the preceding one we were chiefly concerned with the general position of the ovule in the ovary, at the side of which a young stamen (*st*) is represented, and the figure was also intended to illustrate the delicate and feathery structure of the stigma.

The actual process of fertilisation in the receptive ovule is represented (after Kny) in the case of the ovule of the Pansy (*Viola tricolor*). The ovule has been detached from the ovary wall at the point *p*, the **placental region**, which produced the ovule. The vascular bundle (*r*) of the placenta is continued for the purposes of nutrition into the base of the ovule. The tissue in which it terminates in the ovule has been termed the **chalaza**. The nucellus is enclosed by the **inner integument** (*JJ*) and the **outer integument** (*AJ*), which however leave the small canal or micropyle open. The pollen tube has penetrated into this passage, has grown through the apex of the nucellus (*KW*) towards a large and almost cylindrical cell, the **embryo sac**.

The embryo sac has become developed from one of the cells of the nucellus by considerable enlargement at the expense of the neighbouring cells, which broke down into mucilage. It contains at the commencement a large nucleus, which divides, the daughter-nuclei progressing to the two extremities of the embryo sac. There, by repeated division, a group of four nuclei results. A later stage of development is shown in the case of the embryo sac of *Polygonum divaricatum* (Fig. 33, after Strasburger). Two of the cells (*s*) at the apex of the embryo sac exhibit each a large vacuole and towards the apex a nucleus; the third one (*e*) has the nucleus at the base and the vacuole at the apex.

It is only this latter cell, the **egg-cell**, which gives rise after fertilisation to the young plant (*embryo*). The other two cells, characterised by the striped appearance they represent, have been termed the **synergidæ**. The group of three cells is called the **egg apparatus**.

The cells at the opposite end of the embryo sac are sur-

FRUITS AND SEEDS.

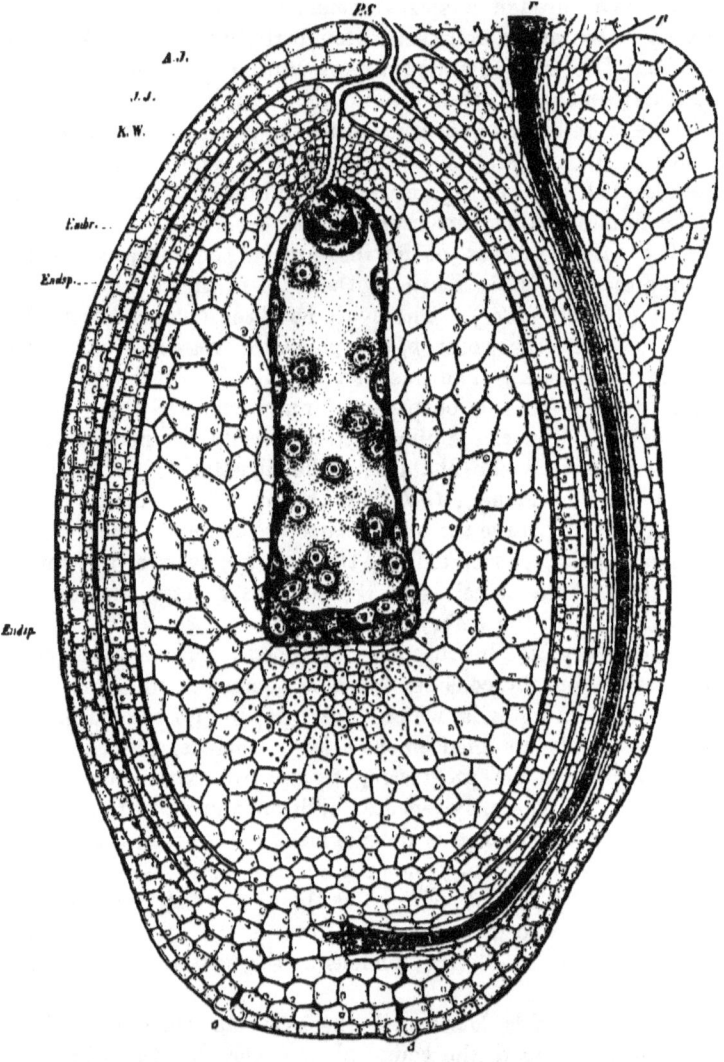

FIG. 52.—LONGITUDINAL SECTION THROUGH AN OVULE OF *Viola tricolor* (*after* KNY).
PS, pollen tube ; *AJ*, outer integument ; *JJ*, inner integument ; *KW*, nucellus ; *Emb*, embryo ; *Endsp*, endosperm ; *p*, placenta ; *r*, fibro-vascular bundle ; *s*, stomata.

rounded by a thin cell-wall, and differ in this respect from the cells of the egg apparatus. But we have stated that there were four cells at each end, and we have so far only spoken of three; there remains, therefore, one nucleus at either end. These will, however, be noticed to move towards the middle of the embryo sac, and there to fuse into a new nucleus the secondary nucleus of the embryo sac.

Having reached this stage of development, the constantly enlarging embryo sac is now receptive, and the pollen tube makes its way through the nucellus down to the embryo sac. In some cases the pollen tube has less resistance to overcome, as the tissue of the nucellus lying above the embryo sac becomes disorganised (*Labiatæ*), while the wall of the embryo sac, lying directly below the micropyle, breaks down or is broken down by the synergidæ.

Before reaching the entrance to the micropyle, the developing pollen tube may find certain arrangements which make its progress more easy. Thus we find ridges, or papillæ, or glandular excrescences, or specially loose tissues in the style which lead to the ovules, and are therefore able to direct the course of the growing pollen tube. Every ovule of the ovary requires a special pollen tube for its fertilisation, and as there are enormous numbers of small ovules in some ovaries (*Orchidaceæ, Ericaceæ, Gesneriaceæ*), we can well perceive how it is that sometimes bundles of innumerable pollen tubes will be met with in the style, visible to the naked eye as a thread of delicate silky hairs.

During this period of active growth the pollen tube is thin-walled, and its basal portion, which is next to the pollen grain, is transparent, because all the contents are collected near the apex. As soon as the pollen tube enters the micropyle, it begins to swell up and becomes mucilaginous. This process of swelling up and softening of tissues, as we have seen above, is no uncommon occurrence in the case of the nucellus, or indeed of the wall of the embryo sac. It enables the contents of the pollen tube to pass more easily into the egg-cell lying at the apex of the embryo sac.

The transference of the fertilising contents of the pollen tube must take place by a diffusion through the softened membrane, as a perforation or disruption of the cell-wall has not been observed. One can only see that the contents of the pollen tube become squeezed towards the embryo sac by a constricting of the micropyle, and that the tube becomes firmly applied to one of the synergidæ. Then the nucleus and vacuole of this cell disappear; it and its sister-cell (Fig. 33, *s, s*) become dim and granular, lose their shape, and are ultimately seen attached to the egg-cell as colourless and irregular viscid masses.

This completes the act of fertilisation, the immediate effect of which is that the egg-cell (*e*), having temporarily two nuclei (which afterwards fuse), surrounds itself with a delicate membrane.

This stage of the development is shown in Fig. 32. *Embr* represents the fertilised egg-cell, and we notice that during or shortly after fertilisation several changes have taken place in the embryo sac. Before fertilisation, as is mentioned above, a fusion of one of the nuclei of the micropylar end with a nucleus from the antipodal end has given rise to a single new nucleus near the centre of this embryo sac. Now by repeated divisions of this nucleus, a large number of nuclei (*Endsp*) have been formed, which

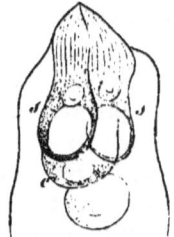

Fig. 33.

are chiefly found in the layer of protoplasm lining the inner wall of the embryo sac. Here each nucleus becomes the centre of a new cell, and by further division the embryo sac becomes entirely filled by a new cellular tissue. This tissue, which fills the entire embryo sac when the latter has completed its growth, is called the **albumen** or **endosperm**. It is the tissue in which are stored up the reserve substances which have to nourish the young embryo during its further growth, and it forms the greater part of some seeds. In the case of the wheat-grain and other grass-seeds, we see the small embryo attached to one end of the seed, and the rest of the tissue filling up the former embryo sac is the starchy endosperm, very much more considerable in bulk than the embryo. If we imagine the centre of this mealy endosperm to be hollow, and not filled up with cells, we should have a picture of what actually occurs in the cocoa-nut. The hollow centre of the cocoa-nut, in which the milk is contained, is nothing else than the central cavity of the embryo sac, which has grown to so enormous a size, and becomes only partially filled with endosperm cells, forming the edible coat of the nut.

But all seeds do not possess a large mass of albumen at the side of or around the embryo; many seeds are exalbuminous. In such cases the development of the embryo greatly exceeds the growth of any of the other parts of the ovule; the cells of the endosperm become absorbed again, and are made use of by the developing embryo, which grows so large that it fills up the entire cavity within the two integuments. We then get a seed of the same kind as that of the bean. In this case the former integuments of the ovule have formed the hard seed-coat (testa), and the latter encloses the embryo, which was formerly represented by a single cell at the apex of the embryo sac. The dark oval patch on the glistening testa is the point at which the stalk of the ovule (funicle) was attached, fastening the young seed to the placenta. This mark is termed the **hilum** of the seed.

The main portion of the seed consists of the two large seed-leaves or cotyledons, which will be found attached one to the other at the base, the attachment being continued into a small papilla **(the radicle)** at one end, while between the cotyledons is enclosed the embryonic shoot **(plumule)** with its rudimentary leaves. That these parts are really the young organs of the plant can best be seen on the germination of the seed. The seedling then bursts through the testa, the young root apex growing down into the soil. Then the two cotyledons expand as succulent leaves, and between them grows up the young shoot. As the latter increases in size the cotyledons become poorer in contents, wrinkled, and ultimately shrivel up and fall off; they have performed their duty in supplying the young root and shoot with the food material with which they were filled. The cotyledons, therefore, have the same function in the exalbuminous seeds which the endosperm performs in the albuminous seeds, namely, to nourish the seedling during the early stages of its growth.

The stimulus which the process of fertilisation imparts to the ovule, and which results in the development of the seed, is also transmitted to the tissues of the ovary wall. This is best seen in the case of the flowers of pears and apples, where we can distinguish a very few days after the flowering, with a considerable degree of certainty, which of the flowers have been fertilised. In these the flowering stalk is much more turgid and begins to thicken very soon.

With the progressive development of the seed changes also take place in the carpels which bear them, so that the green ovary becomes changed into the ripe fruit. The wall of the ovary often becomes differentiated into three layers: the outermost layer (*epicarp*), a median one (*mesocarp*), and an inner one (*endocarp*). If they all three become hard and dry, the fruit is termed a dry fruit (hazel-nut, pod of pea, or capsules of lily and poppy). If, however, one or more layers become fleshy, then we speak of succulent fruits (gooseberry, fruit of the melon, drupe of cherry and plum). The cherry-stone is not the seed of the cherry, but the hard inner portion of the ovarian wall, and encloses the actual seed. It is also important to know that the fruit of the pear and apple is what is termed a false fruit or pseudo-carp, in the formation of which the axis of the flower or receptacle plays a part. In the case of the strawberry, too, we really eat the fleshy receptacle upon which the actual fruits are borne as numerous small nutlets. In the case of the raspberry and blackberry, however, we are dealing with a number of true succulent fruits which are crowded together on a common receptacle. In the case of the pine-apple, mulberry, and fig, the entire inflorescence becomes sweet and succulent. The spiny coat surrounding the nuts of the Spanish chestnut is not the ovary wall, but an involucre or cupule made up of four bracts, and of the same nature as the cup of the acorn.

§ 45. How can the formation of the fruit be influenced by different methods of cultivation?

At the end of the last chapter we dealt with the mode of formation of fruits and of fruit-like structures (*pseudo-carps*), which in general parlance are termed fruits. Now we shall discuss the processes by means of which the character of the fruits may be changed and improved.

In the case of the fruits used for dessert, it is essential to increase their number, size, appearance, and flavour.

An increase in the number of fruits demands an increase in the number of flower-buds, and as these are generally formed in the preceding season, the horticulturist must extend his operations to that period. Everyday experience teaches us that the period of sexual reproduction does not, as a rule, set in until the vegetative growth (production of leaves) begins to diminish or to cease entirely. The function of the leafy tissue is to form new organic food material from the raw material contained in the soil as inorganic salts and in the air as carbonic acid. The stronger the young embryo, the stronger will be the growing seedling—that is, the active plant. The larger and more numerous the organs and the greater the quantities of raw material which they can assimilate in a given time, the greater will be the number of new assimilating organs which the plant will develop. The longer these organs function, the greater will be the amount of substances assimilated.

The duration of the assimilating period, and the way in which the products are used up, depend partly upon the season in which the vegetative period falls. In the case of plants left to themselves, the germination of the seeds and the development of leaves in trees and shrubs takes place in spring, when the soil is moister, but during which period there is less light and heat than in summer. This combination of circumstances proves to be very beneficial for the development of leaves. On the other hand, the foliage is smaller, the shoots shorter, and the number of lateral branches smaller, and their development less extensive, if the vegetative period is artificially postponed until the summer, or if the spring is abnormally dry and warm. The plant will under such circumstances very soon come to an end of its vegetative period, and begins at once to use the assimilated substances for the period of sexual development.

If, however, the first period is lengthened by artificial shading, copious watering, and rich nutrition, or if a moist spring is followed by a damp, cool, and sunless summer, then the period of leaf production will never really come to an end. In such a case the plant uses the greater portion of the assimilated matter for the further development of its leaves, *i.e.*, for the increase of its leafy tissues, and there is generally no opportunity for storing up the large amount of reserve material necessary for the production of flowers. In the case of our trees, we may also observe that the formation of fruits is not commenced in the earlier years of their growth, when the trees are occupied with the development of their crown. Oaks generally attain an age of fifty years, pines forty years, firs twenty years, and our fruit-trees, with artificial aid, twelve or fifteen years before the development of new branches has so far abated as to allow the formation of short flowering shoots to take place. Only when this youthful period of life is over does the apical growth of the branches cease at the proper period and sufficient nutriment is stored up in the axis. The greater the number of leaves on each shoot, the larger will be the amount of food matter stored up. In the case of plants, therefore, in which we desire to increase the number of flowering buds, our method of cultivation must aim at producing as large a number of leaves as possible, and at giving them a sufficient time for assimilation. But we must also ensure a period of rest at the right time.

These conclusions, drawn from daily experience, will enable us now to criticise the treatment of our dwarf and trained fruit-trees. We can now see why, in many cases where the branches are well trained, few or no flower-buds make their appearance. The careless pruning of the branches in the summer, and the too frequent pinching off of the tips, stimulate many of the buds which should not develop till next year to grow out prematurely. The tree exhausts its forces, therefore, in the production of new shoots without retaining the leaves long enough to produce the requisite amount of food material. We have already touched upon this point in speaking of the treatment of shoots.

To annuals and biennials the same rules apply as to shrubs and trees, but we are not able here to correct the faults of one year in a subsequent season. We must remember here from the beginning that a very large number of flowers can only be attained by the continuous functioning of a very large amount of foliage. Assuming therefore that the soil is rich in nutritive salts, as should always be the case in successful horticulture, we must aim at producing as early a germination of the seeds as possible. Furthermore, the development of the leaves must be helped on by quick-acting manures containing potassium and nitrates, which should be applied as top-dressing. But above all, there should be a copious supply of water to aid in the elongation of the cells. These artificial aids should gradually be decreased as the hot season approaches. The more powerful effect of the sun and the decreasing amounts of water will diminish and ultimately stop the production of new leaves, and give the plant the opportunity of forming flowering buds and help on the ripening of seeds and fruits.

It depends entirely upon the specific peculiarities of the plant and upon the aim of the gardener as to whether any artificial aid should be resorted to after the flowering season. If the production of seeds is the chief aim of cultivation, it seems best to leave the plants to themselves and to let the food material which has been formed by the leaves in the preceding vegetative period, but which has been stored up in the stem, pass over into the ripening seeds. This passage from the older to the younger portions of the plant takes place very slowly, the leaves becoming gradually poorer in food matter and turning yellow. Manuring during this period is directly injurious.

If, however, it is the aim of the horticulturist to produce luscious and edible fruit, in which either the ovary wall (*pericarp*) or the receptacle is to become succulent and sweet, it will be very beneficial to accelerate the natural processes by artificial means. For in this case it is not directly the sexual organs which have to be nourished by the already formed food material, but it is the growth of part of the vegetative structure of the plant which has to be promoted, and therefore those stimuli may be resorted to which tend to produce foliage leaves.

In the case of dessert fruits, the full development of the succulent tissues, their tenderness, lusciousness, and sweetness are objects of attainment. It has so far not been proved that a fruit will develop many new cells during its growth to maturity. This may take place to a limited extent; but from the fact that large fruits do not contain many more cell-layers than small fruits of the same variety, we may assume that the differences in fruits are mainly due to a difference in the development of the cells. The size of the cells, however, and the amount of their contents are exceedingly variable. The development of the fruits can therefore only be accelerated by such means as increase the elongation of the cells, and cause weak cells to develop healthily. In horticultural practice, therefore, repeated watering and moderate application of liquid manure is resorted to.

The effectiveness of this procedure seems to be greater the more the organ which develops into the fruit resembles a vegetative axis. Liquid manure (in small quantities), or water alone, prove therefore especially effective in the swelling of apples and pears, figs, pine-apples, and strawberries. The apple may be looked upon as a cup-shaped apex of a stem, in which the carpels are sunk. The succulent portion of the pear or apple is, therefore, mainly an excessively succulent cortical tissue, as can best be seen in those cases in which the fruit is continued on one side down the fruit-stalk. In the case of the strawberry, the succulent axis is not hollow, but forms a conical protuberance, and in its development we may sometimes see the bad effects of too copious a supply of water or nutritive salts. Many flowers remain sterile, and the smaller number produce enormous fruits, which often become hollow at their centre. The elongation of the cells is so excessive that a considerable tension exists at the centre, and the cells separate one from another, as is the case in other axial structures under similar circumstances (turnips and potatoes).

In the case of stone-fruits, a judicious supply of water will also aid in the swelling of the fruits, but here considerably more care is needed in selecting the proper time for such operations, as otherwise the fruits are liable to be thrown off. Especially in the case of peaches, apricots, and plums it seems desirable to abstain from the application of liquid manure or from watering until the fruits have attained about one-half of their normal size, and until the sclerenchymatous cells which form the stone are sufficiently developed. During the period in which the stone is formed the fruit does not increase much in size, which seems to indicate that the fruit does not during that time require so much water. It is just at this period that the fruits drop if the plants are copiously watered. One should also guard against beginning watering too suddenly in case the trees have been passing through a considerable period of drought, which may have prematurely coloured the fruit. A sudden and copious supply of water may cause a total fall of the fruits; the water supply has to be begun gradually, so as to get the tree used to the more active transpiration which this necessitates.

We may here add, that the usual treatment of plants which have been suffering from drought is a wrong one; as a rule, they are copiously watered, shaded, and, where it is possible, they are placed in a saturated atmosphere. One forgets that the plant has become unaccustomed during the period of drought to any great transpiration, and this must gradually increase before the water can be properly absorbed. The transpiration can, however, be increased by placing the plant after the first watering in a light place and in a dry atmosphere. We can easily detect in a plant placed on a balance that the amount of transpiration gradually increases and the plant will recover, whereas a plant suddenly swamped with water and placed in a dark damp place will recover very slowly indeed, or may die altogether.

§ 46. **What are the conditions governing the production of seeds?**

The ovules become changed by the act of fertilisation, *i.e.*, after the contact of the pollen tube with the egg-cell, and develop into seeds. Without fertilisation no seeds are ever developed; but a fruit, especially a pseudo-carp, formed by a succulent axis, may develop without any fertilisation having taken place. Thus pears may become fully formed by abnormal development, and will be found to be without a core, but edible.

Fertilisation is dependent upon pollination, *i.e.*, upon the transference of pollen grains to the stigma of a similar flower. Seeing that the gardener has in many cases, in order to obtain seeds, to proceed to artificial pollination by transferring, with the aid of a fine brush, the pollen to the stigmatic surface, we may point out that there are many hindrances to this method of procedure. In nature, too, we may often observe that the pollen from opening stamens may fall upon the stigma of the same flower (self-pollination) but remains without effect. This is the case in so called dichogamous flowers, in which the two kinds of sexual organs ripen at different times. In some cases the anthers ripen and disperse their pollen before the stigma of the flower is receptive. Such plants are termed **protandrous** (Geraniaceæ, Malvaceæ, Compositæ, &c.), and fertilisation can only be effected with the aid of a second, younger flower, by either natural or artificial means. In the reverse case, namely, in that of the **protogynous** flowers, the stigma which ripens before the stamens requires to be fertilised with pollen from an older flower, because its stamens are as yet not fully developed (some Gramineæ). But even in the case of both sexual organs developing at the same time, self-pollination may often be difficult, if not impossible, because the stamens are not long enough to reach the stigma. This is the case with **dimorphic** flowers. In this case a plant will develop hermaphrodite flowers of two kinds (Primrose). Some specimens have a long style, so that the stigma will reach to the top of the corolla tube (*pin-eyed*). The stamens, however, will be short, and remain hidden within the tube of the corolla. In other specimens the reverse will be the case (*heterostyly*).

Nature seems here to point out the way in which fertilisation is to be effected. The flowers with long stamens are intended to pollinate the long-styled forms. That this course of pollination is attained with good results has been proved experimentally by Darwin. He found that the production of seed is small, or sometimes even nil, when one (short or long-styled) form is pollinated with its own pollen, or with pollen from a similar flower, but that a large production of seed results from the pollination of a long-styled form with pollen from a short-styled form.[1] We might add to these instances of heterostyly a description of other mechanisms of flowers which render it necessary for a plant to have special carriers or distributors of pollen (fertilising agents). We need only refer to the compact pollen masses (*pollinia*) of the Orchids, which could never reach the stigma of their own accord, and to the monœcious and diœcious plants. In the case of the latter, *e.g.*, the Conifers, Willows, Poplars, Oaks, &c., the wind will transport the pollen from flower to flower. Such flowers are therefore termed **anemophilous**, in contradistinction to those in which insects involuntarily, no doubt, perform the pollination (**entomophilous**). This is the case in the Orchidaceæ, Aristolochiæ, and many Papilionaceæ and Labiatæ. In some few water-plants (*Vallisneria, Ceratophyllum*) the water carries the pollen from one flower to the stigma of another one (**hydrophilous**).

Especially in the case of entomophilous flowers do we find the most curious mechanisms for the employment of insects for the purpose of pollination; but we must content ourselves with a single instance.

In the year 1885, the Society for Acclimatising introduced into its gardens in Canterbury, New Zealand, the first humble-bees. In the next year these insects had considerably increased in numbers, and it was then found that the common red clover, which had been introduced into that country, but had formerly only produced very few seeds, now yielded very good seed-crops. The humble-bees, which in this country fertilise the flowers, had till then been missing.

Bees and humble-bees play a very important part in the fertilisation of flowers, carrying the pollen from flower to flower by means of their hairy bodies. They sometimes, however, bore holes into flowers, and then obtain the nectar or honey without pollinating the flower. This is the case in the

---

[1] The pollination of heteromorphic forms with pollen from a similar form of flower has been designated by Darwin an "illegitimate union." The legitimate union consists of the pollination of flowers with pollen from a dissimilar form. Such a beneficial result of cross-fertilisation has not been universally proved—indeed, in some cases self-fertilisation results in a plentiful crop of seeds; yet gardeners will do well in all cases to transfer pollen of one flower on to the stigma of another flower.

Columbine (*Aquilegia*), the spurs of which are perforated by humble-bees. The same is done by them in the flowers of Weigelia, whereas the honey-bee enters the flower. In the different species of Orchis, the spur will often be found ripped open, and in *Tropæolum majus* there are often two or more holes in the spur, made by these insects.

Water-plants, too, are fertilised by insects, e.g., *Nymphæa, Nuphar, Utricularia, Aldrovandia, Hottonia, Stratiotes, Trapa*. This, however, is only the case when the brilliantly coloured petals are expanded above the water. If the flowering stalks remain short, so that the flowers remain below the water, then **cleistogamy** is resorted to. In this case small rudimentary hermaphrodite flowers are formed which do not open, and must therefore of needs be self-fertilised. In such cases self-fertilisation is generally very effective, and in some cases these cleistogamous flowers are the only fertile ones, while the conspicuously coloured and open flowers remain sterile. Among land plants, the Violet is the most well-known example of cleistogamy. Of our wild flowers, the Dead Nettle (*Lamium*), the Rest-harrow (*Ononis*), the Toad-flax (*Linaria*), the Wood-sorrel (*Oxalis*), and others, are provided with cleistogamous flowers, which are not only smaller, but have rudimentary petals devoid of scent and of nectaries.

Attention must be paid to all these conditions for any success in one of the chief lines of horticultural enterprise, namely, in the development of new varieties by crossing. In spite of many successes, we must admit that the latter are due more to chance than to any carefully planned method. It is true that in this field science has as yet only a few data to go upon and no general principles to lay down, and it might derive much benefit from the observations of horticulturists. In horticultural practice it is generally not possible to take all the precautions of removing the stamens from the flower which is to be fertilised, to protect the flower after fertilisation, and to keep a detailed account of all the cross-fertilisation performed, and very often an assumption of certain processes having taken place has to do duty for actual observation. Of course this leads to mistakes. If we acted quite honestly, we ought often to assert that the origin of certain cultivated forms is

Q

doubtful. The greatest certainty exists, of course, in those cases in which specialists are occupied with a single genus or species. Then the repetition of former results may take the place of experiments performed with all the precautions which ought always to be taken when a cross-fertilisation can only be attempted once.

The Orchids seem to be a group especially well adapted for such experimental work. They are now-a-days the fashionable flowers, and their fertilisation is carried out by gardeners with great care; besides which, their flowers can only be fertilised by very special insects, which do not exist in this country. Fertilisation by wind or insects while the plants were under cultivation in Europe has only been recorded for *Phajus grandifolius*, *Aërides affine*, *Vanda Roxburghii*, *Lælia cinnabarina*, *Cypripedium Schlimii*, *javanicum*, *virens*, *Bullenianum*. The signs which indicate that a flower is really fertilised do not always appear in the same way or after the same lapse of time. Gardeners must therefore not be too rash in drawing conclusions from the non-appearance of certain signs. The most well known of these is the swelling of the ovary, which takes place in some cases very rapidly, as has been observed in *Oncidium* (*Papilio Forbesii*), *Odontoglossum*, *Lælia*, *Cattleya*, *Phalænopsis*, &c. Other orchids close their flowers the day after the fertilisation has been effected. *Cypripedium* remains fresh for a considerable period, and its ovary shows no sign of swelling.

On the other hand, it must be pointed out that changes which generally take place after pollination are not always a sure sign that fertilisation has been effected. Of these, the most obvious one is the growth of the pollen tube down the style of another species or genus. Strasburger, for instance, mentions that the pollen tube from a pollen grain of *Lathyrus montanus* grew down into the ovarian cavity of *Convallaria latifolia*, while tubes from the pollen of *Agapanthus* grew deep down into the style of *Achimenes grandiflora* (but not vice versa). The pollen tubes of *Fritillaria persica* are said to grow into the ovarian cavity of orchids, and to cause the development and swelling of its ovules.

We have no rules of general application for hybridisation.

We can at present only say that hybridisation is more easily effected the more nearly allied the parent plants are, whether varieties of the same species or species of the same genus. We might explain this by saying that the substances contained in the pollen tube, and which have to act upon the egg-cell, will be most similar in the case of kindred forms, and will therefore more easily combine. But we have all sorts of curious exceptions. Thus it has been observed that the pollen of *Orchis Morio* produces no tubes on the stigma of *Orchis fusca*, while the pollen tubes of *Orchis fusca* not only penetrate to the ovules of *Orchis Morio*, but effect a fertilisation of the egg-cell.

With regard to the fertility of hybrids, it was formerly held that their vegetative vigour was accompanied by sexual weakness. In some cases the stamens, in others the ovules, would be incompletely developed. Their fertility was, however, said to increase if they were fertilised by one of the parent forms and became again more like it. These statements must not, however, be regarded in the light of a fixed rule, for they hold good, too, in the case of the production of new varieties—that is, when flowers of the same species are fertilised with pollen of the same species.

Our entire method of cultivation tends to increase the sterility of our cultivated plants.

This is caused by the endeavour to grow vigorous plants, in which the vegetative growth is stimulated by abundant water and food supply, until the functioning powers of the various organs are taxed to their utmost limit. Our cultivated plants exhibit the most luxurious foliage, the most rapid growth, the most succulent tissues, the largest flowers, and the most juicy fruits. But this all entails a retrogression of the sexual reproduction, as is shown by the proliferation and the doubling of flowers, in both of which processes the stamens and carpels are transformed into sterile leaves.

Such tendencies are inherited, and are manifest in the offspring of all the most luxuriant of our cultivated plants, without having been caused in any way by hybridisation.

§ 47. How should the ripe seed be treated?

The product of fertilisation, the seed, contains, as chief constituent, the young plant or **embryo**. The latter is generally so far developed, that it shows the embryonic root or **radicle** and the young shoot or **plumule**. Nature has also provided for the early nutrition of the plant before the root begins to function. The seed is for this purpose provided with a mass of food material, which only requires to become dissolved in order that it may be at once made use of by the embryo. This food material, which is stored up chiefly as starch or oil, is either contained in the seed-leaves or **cotyledons**, if the embryo alone fills the seed, or in a special parenchymatous tissue, the **endosperm** or albumen, in which case the embryo is generally small as compared with the size of the seed. In this latter case special arrangements are made, by which the absorption of the dissolved food material by the embryo is as easily effected as if the food material were stored up in the cotyledons.

Besides their economic importance as food materials for man, the seeds must be considered also with regard to the future production of new plants. The chief value of seeds indeed is the reproduction of the species to which they belong. This reproduction is sometimes effected immediately after the ripening of the seeds, but generally they pass through a resting period of some length. We must, therefore, first consider the **resting** and then the **germinating seed**.

The fully mature and garnered seed is by no means lifeless. A number of insufficiently known changes take place within it, which cause it to give off water and carbonic acid, and also to change its colour. We must also assume that during the resting period ferments are formed which cause the rapid solution of the food material during germination. This may be gathered from the fact that seeds of the same kind and of the same crop will under similar conditions require a longer time for germination, the further off they are from the normal time of germination.

The chief condition for securing germination is the requisite supply of water, besides an increase in temperature and in the supply of oxygen. If, therefore, we wish to keep seeds healthy during their period of rest, we must avoid dampness. The latter will be the less harmful the lower the temperature is and the less oxygen there is present. If seeds are not properly attended to, they become musty and ultimately mouldy. If we have the choice between a warm place liable to considerable variations of temperature and a uniformly cool place, we ought always to choose the latter, regardless of the fact that the seeds might be exposed to a frost. Most seeds of our cultivated plants will resist comparatively low temperatures if they are kept completely dry The storehouses for seeds should be **dry, cool, and airy**.

During the process of germination three definite periods may be recognised. First there is the **swelling** of the seed. This process may be looked upon as a more or less mechanical one, and is accompanied by a rise of temperature (due perhaps to the condensation of moisture). This absorption of water brings about the second phase, the **solution of the food material**, which is a chemical process caused by the action of certain ferments, and causing the third change to take place, namely, the increase and **expansion of the embryo**.

The rapidity of the germination depends mainly upon the rapidity with which the water is taken up by the testa. In some plants the epidermal cells of the testa, and sometimes also the adjoining cells, have a palisade-like arrangement, and take up water so rapidly that in a few hours they form a mucilaginous coating to the seed. This is the case with the seeds of pears and quinces, and with linseed. In other cases, the structure of the testa is so firm, and the cuticle covering its epidermal cells renders it so impervious to water, that perfectly sound seeds may lie for years in a damp soil without germinating. This is often the case in leguminous plants (*Robinia*, Clover, &c.). In the case of the latter it has been shown that the palisade-like epidermal cells of the testa, which contain the colouring matter of the seed, are so impervious to water, that, after lying in water for several years, the seeds were still alive and had not germinated.

Sometimes the loss which occurs owing to the non-germination of a considerable percentage of the seeds may be avoided by sowing the seeds very soon after they have ripened. This may be done with advantage in the case of Taxus, Ilex, Magnolia, and Auricula. In the case of strawberries the fruits should be dried in the sun and then be crushed by rubbing them between the hands. They should then be sowed directly, after having been mixed with a little finely sifted earth. If seeds are to be germinated which usually require a considerable time until their testa is soaked, it is best to treat the seeds by some mechanical process before sowing them. Such seeds may be mixed with fine sand in a bag, and then be rubbed or pounded. Experiments with different leguminous plants showed that by such means the absorptive capacity might be increased as much as 30 per cent. In the case of seeds with still harder seed-coats a rougher treatment may be employed. In some cases, where larger and more valuable seeds are to be dealt with, the testa may be opened up by one or more cuts.

In nature, the seeds are sown immediately after they have ripened, and in the case of those which are surrounded by a stony envelope, and which do not germinate in the same year, they are very often surrounded with a succulent layer (Cherries, &c.). In this case it is obvious that the permanently moist covering to the seed will prevent them becoming hardened, as they do when they are dried. We must, where possible, imitate this mode of procedure in horticulture, and sow the seeds with their succulent covering. But in many cases this is not possible, as the seeds have to be cleaned and dried for purposes of sale and transport. We might, however, at least place the seeds in the soil in the autumn, so that they should make use of the moisture of the winter. But this is not always possible on account of the attacks of mice and other animals upon the seeds, and other methods such as the following are usually employed.

The seeds are embedded in sand or sandy soil. They are placed in boxes, baskets, barrels, or pots, according to circumstances, and placed in the cellar or kept in the open, according to local climatic conditions. It is, however, essential in all cases that the receptacle should be well drained, and that a considerable layer of sand should be placed at the bottom, covering a layer of potsherds or cinders. Then the seeds are spread out singly over the sand and covered with another layer of sand. The number of layers of seeds which may be placed one above the other, and the thickness of the intervening layer of sand, depends upon the porosity, and therefore upon the coarseness of the sand. The more porous the latter is and the larger the seeds, the greater the number of layers which may be arranged one above the other. The layer of sand should never exceed three inches, and more than six layers of seeds should not be piled up, as one of the chief requisites, namely, thorough ventilation of the soil, can then not be effectively secured. The moisture must not be too great, but constant. The time at which to begin this stratification depends upon the time the seed requires for germinating. The more rapidly the latter takes place, the later must this process be undertaken, as the radicles must not have grown out much when the seeds are transferred to the open. The effect of this process is heightened by immersing the seeds in water before they are laid in layers, or by keeping the temperature up. The cellar is usually the best place for this process. If a very large quantity of hard seeds which require to be moistened for a considerable time (Ash and Sycamore) are to be dealt with, they may be placed in the open in pits or mounds in a similar manner.

It is, however, still an open question whether fresh seeds are always the best for sowing. In most cases this will be the case; indeed, in some which soon lose their vitality (Willows, Poplars, Elms) it will be absolutely necessary; but we find exceptions to this rule. Many practical gardeners, indeed, affirm that old seeds of melons and cucumbers do not run to leaf so much as fresh ones. This might be readily explained; for we know that plants which have always been copiously supplied with water transpire more from their leaf surface and have a more vigorous development of leaf surface. It is also known that an artificial drying of seed-potatoes or onions causes the production of short-shooted plants, which more readily form tubers and bulbs, and this might also be the case with the seeds of the melon. If seeds are kept for a long time in a dry atmosphere and with little oxygen, they retain their vitality for a long time. Of a number of pine seeds which were kept in a glass jar in a room for five years, 32 per cent. were germinated; after seven years 12 per cent. had sufficient vitality to germinate. Of seeds stored in the same way the following percentages of germination were obtained:—Red clover after twelve years, 10.5 per cent.; peas after ten years, 47.7 per cent.; linseed after six years, 49 per cent.; after eleven years, 3 per cent.

It is often stated that as the production of weakly offspring increases with the age of the seeds, it is advisable to strengthen their vitality by immersing them in dilute chloric, hydrochloric, or some other acids. But the efficacy of this process has not yet been established by experiment. It is possible, however, that if the testa does not readily absorb water, the treatment with acid might loosen the cells and effect a more rapid imbibition.

The **second stage** of germination, the solution of the food material, requires, besides the water supply, above all things an increase in the supply of oxygen. The seeds may even dispense with some of the water—that is, their tissues need not yet be saturated with water—for the commencement of the vegetative changes to take place. If liquid water is not available at the commencement, the seeds may condense and absorb water from the atmosphere, and even absorb hydrogen, nitrogen, and oxygen, like some porous substances. Grain which has been swelled can take up more oxygen out of the atmosphere than nitrogen, and the giving off of carbonic acid takes place so actively, that more carbonic acid is given off than can be accounted for by the amount of oxygen taken in. This indicates that **internal processes of oxidation** (respiration) are carried on in germinating seeds. This oxidation is accompanied by an evolution of heat, and the latter again increases the solution of the reserve substances.

These now make their way to the seedling. If they are contained in the cotyledons or seed-leaves, they will be carried by the young vascular bundles to the root and stem apices. If they are stored up outside the embryo in the form of endosperm, the embryo will be provided with special absorptive cells. In our grains (wheat, maize, &c.)

the embryo is provided with a shield-shaped mass of cells, the **scutellum**, which is applied to the starch-containing tissue. The outer cells of the scutellum have finger-like processes, and constitute the absorptive cells, taking up the food material and passing it on to the embryo.

During this period of germination we must take care that there should never be a dearth of oxygen or a superfluity of carbonic acid. No gas is more injurious to germination than carbonic acid. If there is an admixture of only a few hundredth parts of carbonic acid to one part of oxygen, the germinating process will cease.

Most of the mistakes in germinating occur from the fact that there is a dearth of oxygen and an overplus of carbonic acid. The direct cause of this is either too thick a covering of soil over the seeds, or a closing of the soil to the free access of oxygen by too continuous watering. There can be no fixed rule about the thickness of the covering layer; it depends partly on the size of the seed, and partly on the porosity of the soil.

The main consideration which will enable any one to determine the depth at which his seeds should be sown must always be this, that the soil is the medium which, in the first place, is to keep the seeds sufficiently moist for germination; secondly, and only in those cases in which the seedling will continue its development on the spot where it has germinated, the soil must be looked upon as essential for the fixing of the plant. In horticultural practice, where the young plants are **pricked out**, the covering of the seeds with soil is not essential, if we only keep the seed-pans covered with a pane of glass, or sow the seeds in an open bed in a forcing-pit. But it is different in the case of seeds sown in the open, where the dry winds or hot spring weather may easily cause a temporary or a fatal stoppage of growth. These evils must be overcome by covering the seed with sufficient soil to prevent their becoming dry, but without cutting off the supply of air.

If one is forced to use sunny beds for seedlings, it is better to retain the necessary moisture in the soil by covering it with loose straw, pine branches, &c., than by continual watering. Even if the seeds are not sown too deeply, they may **not germinate if the soil becomes caked** by repeated wetting and drying, &c. Such caking of the soil prevents the access of air to the seed at the time when the greatest amount of chemical change is taking place within them, and the air actually contained in their tissues is not sufficient for any considerable length of time.

How essential the air contained in the tissues of the seed is for germination, may be gathered from experiments which were made with the seeds of turnips. These were placed under the receiver of an air-pump, and the air they contained was pumped out and replaced by water. The seeds took up 71.13 per cent. of water, but only 30 per cent. of them germinated, while control-experiments showed that 90 per cent. of the normal seeds germinated. These latter also developed much more rapidly. In adult plants, too, death may be brought about by extracting the air from the intercellular spaces where it is contained, and replacing it by water. The plants then have a transparent appearance.

A disturbing effect is often caused in seeds which have been swelled, by an interruption of the process of germination by drought. The injuries thus caused vary according to the nature of the plant and the stage of germination at which the disturbance occurs. Speaking generally, the seeds of Monocotyledons seem to suffer less than those of Dicotyledons. The naked grains, like those of the wheat or rye, are especially resistant. Those covered with glumes (bracts), like those of the barley and oat, are more sensitive, and the Indian-corn is especially delicate. Still, all these grains suffer less than the linseed, rape, clover, peas, and beans. Experiments made with a view to studying this phenomenon showed that the cereals are able to withstand even repeated interruptions of their germination by constantly forming new adventitious roots at the base of the old ones, which died off. This faculty is not possessed by most Dicotyledons; the dried-up roots rot away, and their decay affects the adjoining tissue and spreads upwards. Even when decay does not set in, and the seedling gradually recovers, it takes place much more slowly.

Seeds which have once been soaked and have then dried up again will absorb water much more rapidly the second time, but the testa is no longer the same. The increase of the seed during germination may amount to 100 per cent. of its volume, and this means that the testa becomes enormously stretched. If it dries up, the testa shrinks again, and cracks in innumerable places. When the seed becomes moistened again, air and water will have a freer access to the tissues, and the reserve material will become more rapidly dissolved. It will, however, also pass more readily through the cell-walls to the outside, and thus be lost to the embryo.

We see, therefore, that the swelling of seeds previous to the sowing can only be advantageous if we can protect the sown seeds from drought. If this is not possible, then it would be better to leave the seeds to themselves. This rule, that vegetative processes should only be accelerated by copious water supply, if it is possible to continue such a supply, applies also to the **third stage** of germination, and to all stages in the development of vegetative organs. A plant which has been well watered from its youth will give off more water from its leaves than a plant of the same species, and with the same amount of leaf surface, but which has always received only a moderate supply of water. The former will, therefore, fade, and ultimately be injured, under conditions under which a plant used to a smaller water supply and less transpiration can thrive very well. Hence it is that we find that luxurious lawns will suffer more than meagre ones, and luxurious pot-plants more than their small-leaved relations.

It is an agricultural rather than a horticultural practice in some cases to manure the seeds, because it is believed that a stronger seedling would be developed. For this purpose the seeds are either coated with nutritive material or allowed to swell in a solution of concentrated salts. Such a proceeding, where it is not actually harmful, is at least useless. Experiments with turnips, which can surely stand a lot in the way of manures, have shown that ammonium or potassium sulphate, even at a concentration of only 0.5 per cent., is deleterious to germination. Cereals and leguminous plants are also not in the least benefited and, indeed, germinate best in distilled water. Rape-seed alone was able to support a 2 per cent. solution. In the case of seeds which have to lie a long time in the soil before they germinate, the injurious effect of encrusting the seeds is less obvious, because the rain has more time to remove the salts.

INDEX

Absorption by root, 4
Adventitious buds, 169
Aërial roots, 75
Albumin, 124
Aleurone grains, 126
Alkaloids, 32, 126
Aluminium in plants, 36
Ammonium sulphate as manure, 54
Amylodextrin, 122
Amygdalin, 124
Androecium, 214
Apostrophe, 110
Arabin, 123
Arsenic in plants, 37
Asparagin, 125
Asphyxia, 77
Assimilation, 77, 108
Autumn pruning, 139

Barium in plants, 36
Bark scales, 104, 165
Bast, 13
Bast fibres, 96
Bast parenchyma, 12
Bending of shoots, 155
Bleeding of plants, 20
Bone-meal as manure, 55
Boron in plants, 36
Bracts, 221
Bromine in plants, 36
Budding, 183
Bundle sheath, 16

Calcium, function of, in plants, 33
Callus, 149
Callus in cuttings, 173
Callus formed from pith, 190
Calyx, 216

Cambiform cells, 96
Cambium, 98
Cambium ring, 95
Cane-sugar, 123
Capillary attraction, 21
Carbohydrates, 32
Carbon, 31
Carpels, 214
Caryophyllaceae, doubling of, 220
Cell lumen, 17
Cell wall, thickening of, 22
Cells of leaf, 118
Cells of root, 17
Chalaza, 228
Chalky soil, 66
Chili saltpetre as manure, 54
Chlorine, 34
Chlorophyll corpuscles, 18, 130
Chromoplastids, 130
Cleistogamy, 241
Collenchyma, 109
Connective, 215
Copper in plants, 37
Corolla, 216
Corona, 217
Crown grafting, 189
Cuticle, 112
Cuttings, 172
Cystolith, 115
Cytoblast, 127

Dextrose, 123
Diastase, 123
Diclinous flowers, 214
Diffusion, 19
Dimorphic flowers, 239
Dioecious, 214
Dormant buds, 169

INDEX

Embryo, 232, 244
Embryo sac, 228
Emulsin, 124
Endodermis, 16
Endosperm, 231, 244
Epidermis, 5, 95, 112
Epistrophe, 110

Fallow, 52
Ferments, 122
Fibro-vascular bundle, 13
Filament of stamen, 215
Flower, the, 214
Flower, development of, 218
Fluorine in plants, 36
Foliage, damage to, 199
Foliage, removal of, 200
Fruit, pruning for, 142
Fruits, development of, 226
Fungi, 44
Funicle, 232

Germination, 245
Globoid, 126
Glucosides, 123
Glycose, 123
Grafting, 183
Grape-sugar, 123
Growth in thickness, 104
Guano, 55
Guard cells, 113
Gum arabic, 123
Gum tragacanth, 123
Gynæcium, 214
Gypsum, use of, 69

Hadrom, 29
Hairs, 117
Hard bast, 96
Healing processes, 149
Heliotropism, 109
Hermaphrodite flowers, 214
Heterostyly, 239
Hybridisation, 242
Hydrogen, 31
Hydrotropism, 93
Hypochlorin, 130

Integuments, 226
Intercellular spaces, 78, 112
Invertin, 123
Involucre, 221
Iodine in plants, 36
Iron in soil, 34

Jamin's chain, 21

Kainit as manure, 41

Levulose, 123
Layering, 170
Lead in plants, 37
Leaf, the, 103
Leaf, development of, 120
Leaf used for propagating, 201
Leaves, treatment of, 199
Lenticels, 78, 116
Leptom, 15, 29
Leucoplastids, 130
Libriform fibres, 12
Lignification, 12
Lime as manure, 68
Lime, oxalate of, 132
Lithium in plants, 36

Magnesium in plants, 33
Manures, inorganic, 53
Manures, organic, 55
Marls, 66
Marling, 69
Medulla, 95
Medullary rays, 95
Membrane, primary, 11, 22
Meristem, 7, 101
Mesophyll, 112
Micellæ, 19
Micropyle, 226
Microsomata, 129
Monœcious plants, 214
Mycorhiza, 46
Myrosin, 124

Nitrogen, 32, 42
Notching of stems, 158
Nucellus, 226

INDEX

Nucleolus, 17
Nucleus, 17, 127
Nutrition of pot-plants, 70
Nutrition by aërial roots, 75
Nutrition of roots, 30
Nutritive substances, how replaced, 52

OVARY, 215, 226
Ovules, 216, 226
Oxygen, 31
Oxygen, lack of, 77

PALISADE parenchyma, 119
Paracorolla, 217
Parasites, 48
Parenchyma, 12, 109
Pectin substances, 123
Peeling of stems, 161
Pepsin, 123
Perianth, 214
Periblem, 218
Pericambium, 15
Permeability of soil, 61
Phloem, 13
Phosphoric acid, 42, 54
Phosphorus in plants, 32
Pinching, 138
Pit, bordered, 25
Pith, 95
Placenta, 228
Ploughing in of crops, 67
Pneumathodes, 80
Pollen grains, 215
Pollen tube, 227
Pollination, 239
Pollinia, 240
Potassium in soil, 41
Potassium as manure, 54
Pot-plants, nutrition of, 70
Pressure of bark, 104
Pricking out, 249
Primary meristem, 120
Primordial utricle, 17, 110
Procambial strands, 99
Propagation by cuttings, 172
Propagation by layerings, 169

Prosenchyma, 12
Protandrous flowers, 239
Proteids, 124
Protein crystals, 125
Protogynous, 239
Protoplasm, 16
Pruning for fruit, 142
Pruning for wood, 142
Pruning, how to increase effect of, 155

RE-POTTING, 85
Resins, 124
Respiration of leaf, 109
Respiration of roots, 77, 207
Respiratory cavity, 112
Retentiveness of soil, 40
Ringing, 159
Root cap, 9
Root cuttings, 182
Root hairs, 5
Root pressure, 20
Root tip, 7
Root tubercles, 48
Roots, absorptive portion of, 4
Roots, conducting portion of, 10
Roots, how nourished, 37
Roots, respiration of, 61
Roots, structure of, 4
Roots, treatment of, 81, 85
Roots, upward growth of, 79
Rubidium in plants, 36

SALPETRE, Chili, as manure, 54
Salt, common, as manure, 56
Sandy soil, 65
Saprophytes, 44
Scion, influence of, on stock, 194
Sclerenchyma, 102
Scraping of stems, 105, 165
Scutellum of grasses, 249
Secondary thickening, 104
Seeds, formation of, 226
Seeds, manuring of, 251
Seeds, production of, 238
Seeds, treatment of, 244
Self-pollination, 239

INDEX

Shoots, bending of, 155
Shoots, treatment of, 134
Shoots, twisting of, 157
Sieve tubes, 12, 96
Silica in plants, 35
Slitting the bark, 164
Sodium in plants, 35
Soft bast, 13
Soil, improvement of, 65
Soil, permeability of, for air, 61
Soil, power of absorption of, 40
Soil, retentiveness of, 40
Soil, structure of, 39
Sphæro-crystals, 119
Splint wood, 104
Spongy parenchyma, 79, 119
Stable manure, 56, 67
Stamens, 215
Starch sheath, 18
Stem, function of, 105
Sterility, 243
Stigma, 215, 226
Stomata, 78, 113
Storage tissue, 18
Stratification of seeds, 247
Strontium in plants, 36
Style, 226
Sugar sheath, 18
Sulphur in plants, 32
Summer pruning, 137
Superphosphates, 55

TANNIN, 124
Testa, 232
Thylloses, 28, 136
Tongue grafting, 189
Tracheids, 13, 26
Transpiration, 116
Trichomes, 116
Trimethylamin, 126
Trophoplasts, 130
Trophotropism, 93
Turgidity, 20, 28
Twisting of shoots, 157

VACUOLES, 17, 127
Vascular bundles, 10
Vessels, 11

WATER-GLAND, 116
Water-stoma, 116
Watering, theory of, 205
Wild stock, influence of, 194
Wood cells, 11
Wood fibres, 95
Wood parenchyma, 12
Wounds, injuries due to, 135
Wounds, healing of, 149

XANTHOPHYLL, 130
Xylem, 13

ZINC in plants, 37

THE END.

PROPERTY LIBRARY
N. C. State College

Printed by BALLANTYNE, HANSON & CO.
Edinburgh and London